The portrait of Sir Robert Robinson that is printed on p 26 of this volume appears through the courtesy of the Royal Society, London, England. We thank them for allowing us to reproduce this elegant work of art in Michael J. S. Dewar's autobiography.

A Semiempirical Life

A Semiempirical Life

Michael J. S. Dewar

PROFILES, PATHWAYS, AND DREAMS
Autobiographies of Eminent Chemists

Jeffrey I. Seeman, Series Editor

American Chemical Society, Washington, DC 1992

Library of Congress Cataloging-in-Publication Data

Dewar, Michael James Steuart.
  A semiempirical life / Michael J. S. Dewar.

  p.   cm.—(Profiles, pathways, and dreams, ISSN 1047–8329)

  Includes bibliographical references (p.   ) and index.

  ISBN 0–8412–1771–8 (cloth).—ISBN 0–8412–1797–1 (paper)

  1. Dewar, Michael James Steuart. 2. Chemists—
England—Biography. 3. Chemistry, Organic—History—
20th century.

  I. Title. II. Series.

QD22.D35A3   1990
540'.92—dc20
[B]   90–911
                                                                  CIP

Jeffrey I. Seeman, Series Editor

The paper used in this publication meets the minimum requirements of American
National Standard for Information Sciences—Permanence of Paper for Printed Library
Materials, ANSI Z39.48–1984.

∞

Foreword

In 1986, the ACS Books Department accepted for publication a collection of autobiographies of organic chemists, to be published in a single volume. However, the authors were much more prolific than the project's editor, Jeffrey I. Seeman, had anticipated, and under his guidance and encouragement, the project took on a life of its own. The original volume evolved into 22 volumes, and the first volume of Profiles, Pathways, and Dreams: Autobiographies of Eminent Chemists was published in 1990. Unlike the original volume, the series was structured to include chemical scientists in all specialties, not just organic chemistry. Our hope is that those who know the authors will be confirmed in their admiration for them, and that those who do not know them will find these eminent scientists a source of inspiration and encouragement, not only in any scientific endeavors, but also in life.

M. Joan Comstock
Head, Books Department
American Chemical Society

Contributors

We thank the following corporations and Herchel Smith for their generous financial support of the series Profiles, Pathways, and Dreams.

Akzo nv

Bachem Inc.

E. I. du Pont de Nemours
and Company

Duphar B.V.

Eisai Co., Ltd.

Fujisawa Pharmaceutical Co., Ltd.

Hoechst Celanese Corporation

Imperial Chemical Industries PLC

Kao Corporation

Mitsui Petrochemical Industries,
Ltd.

The NutraSweet Company

Organon International B.V.

Pergamon Press PLC

Pfizer Inc.

Philip Morris

Quest International

Sandoz Pharmaceuticals
Corporation

Sankyo Company, Ltd.

Schering–Plough Corporation

Shionogi Research Laboratories,
Shionogi & Co., Ltd.

Herchel Smith

Suntory Institute for Bioorganic
Research

Takasago International
Corporation

Takeda Chemical Industries, Ltd.

Unilever Research U.S., Inc.

Profiles, Pathways, and Dreams

Titles in This Series

About the Editor

JEFFREY I. SEEMAN received his B.S. with high honors in 1967 from the Stevens Institute of Technology in Hoboken, New Jersey, and his Ph.D. in organic chemistry in 1971 from the University of California, Berkeley. Following a two-year staff fellowship at the Laboratory of Chemical Physics of the National Institutes of Health in Bethesda, Maryland, he joined the Philip Morris Research Center in Richmond, Virginia, where he is currently a section leader. In 1983–1984, he enjoyed a sabbatical year at the Dyson Perrins Laboratory in Oxford, England, and claims to have visited more than 90% of the castles in England, Wales, and Scotland.

Seeman's 80 published papers include research in the areas of photochemistry, nicotine and tobacco alkaloid chemistry and synthesis, conformational analysis, pyrolysis chemistry, organotransition metal chemistry, the use of cyclodextrins for chiral recognition, and structure–activity relationships in olfaction. He was a plenary lecturer at the Eighth IUPAC Conference on Physical Organic Chemistry held in Tokyo in 1986 and has been an invited lecturer at numerous scientific meetings and universities. Currently, Seeman serves on the Petroleum Research Fund Advisory Board. He continues to count Nero Wolfe and Archie Goodwin among his best friends.

Contents

Photographs

Preface

"How did you get the idea—and the good fortune—to convince 22 world-famous chemists to write their autobiographies?" This question has been asked of me, in these or similar words, frequently over the past several years. I hope to explain in this preface how the project came about, how the contributors were chosen, what the editorial ground rules were, what was the editorial context in which these scientists wrote their stories, and the answers to related issues. Furthermore, several authors specifically requested that the project's boundary conditions be known.

As I was preparing an article[1] for *Chemical Reviews* on the Curtin–Hammett principle, I became interested in the people who did the work and the human side of the scientific developments. I am a chemist, and I also have a deep appreciation of history, especially in the sense of individual accomplishments. Readers' responses to the historical section of that review encouraged me to take an active interest in the history of chemistry. The concept for Profiles, Pathways, and Dreams resulted from that interest.

My goal for Profiles was to document the development of modern organic chemistry by having individual chemists discuss their roles in this development. Authors were not chosen to represent my choice of the world's "best" organic chemists, as one might choose the "baseball all-star team of the century". Such an attempt would be foolish: Even the selection committees for the Nobel prizes do not make their decisions on such a premise.

The selection criteria were numerous. Each individual had to have made seminal contributions to organic chemistry over a multidecade career. (The average age of the authors is over 70!) Profiles would represent scientists born and professionally productive in different countries. (Chemistry in 13 countries is detailed.) Taken together, these individuals were to have conducted research in nearly all sub-specialties of organic chemistry. Invitations to contribute were based on solicited advice and on recommendations of chemists from five continents, including nearly all of the contributors. The final assemblage was selected entirely and exclusively by me. Not all who were invited chose to participate, and not all who should have been invited could be asked.

A very detailed four-page document was sent to the contributors, in which they were informed that the objectives of the series were

1. to delineate the overall scientific development of organic chemistry during the past 30–40 years, a period during which this field has dramatically changed and matured;

2. to describe the development of specific areas of organic chemistry; to highlight the crucial discoveries and to examine the impact they have had on the continuing development in the field;

3. to focus attention on the research of some of the seminal contributors to organic chemistry; to indicate how their research programs progressed over a 20–40-year period; and

4. to provide a documented source for individuals interested in the hows and whys of the development of modern organic chemistry.

One noted scientist explained his refusal to contribute a volume by saying, in part, that "it is extraordinarily difficult to write in good taste about oneself. Only if one can manage a humorous and light touch does it come off well. Naturally, I would like to place my work in what I consider its true scientific perspective, but . . ."

Each autobiography reflects the author's science, his lifestyle, and the style of his research. Naturally, the volumes are not uniform, although each author attempted to follow the guidelines. "To write in good taste" was not an objective of the series. On the contrary, the authors were specifically requested not to write a review article of their field, but to detail their own research accomplishments. To the extent that this instruction was followed and the result is not "in good taste", then these are criticisms that I, as editor, must bear, not the writer.

As in any project, I have a few regrets. It is truly sad that Egbert Havinga, who wrote one volume, and David Ginsburg, who translated another, died during the development of this project. There have been many rewards, some of which are documented in my personal account of this project, entitled "Extracting the Essence: Adventures of an Editor" published in *CHEMTECH*.[2]

Acknowledgments

I join the entire chemical community in offering each author unbounded thanks. I thank their families and their secretaries for their contributions. Furthermore, I thank numerous chemists for reading and reviewing the autobiographies, for lending photographs, for sharing information, and for providing each of the authors and me the encouragement to proceed in a project that was far more costly in time and energy than any of us had anticipated.

I thank my employer, Philip Morris USA, and J. Charles, R. N. Ferguson, K. Houghton, and W. F. Kuhn, for without their support Profiles, Pathways, and Dreams could not have been. I thank ACS Books, and in particular, Robin Giroux (acquisitions editor), Karen Schools Colson (production manager), Janet Dodd (senior editor), Joan Comstock (department head), and their staff for their hard work, dedication, and support. Each reader no doubt joins me in thanking 24 corporations and Herchel Smith for financial support for the project.

I thank my children, Jonathan and Brooke, for their patience and understanding; remarkably, I have been working on Profiles for more than half of their lives—probably the only half that they can remember! Finally, I again thank all those mentioned and especially my family, friends, colleagues, and the 22 authors for allowing me to share this experience with them.

JEFFREY I. SEEMAN
Philip Morris Research Center
Richmond, VA 23234

November 11, 1990

[1] Seeman, J. I. *Chem. Rev.* **1983**, *83*, 83–134.
[2] Seeman, J. I. *CHEMTECH* **1990**, *20*(2), 86–90.

Editor's Note

To the mind's eye, Michael Dewar appears larger than life, a booming, zooming individual who does not appear cognizant of the extent to which his personality and style affect those around him, nor of the way they respond to him and to his science. He considers himself self-taught and is proud of it: "Because of the way teaching was conducted at Oxford, I was almost entirely self-taught, . . . [which] is probably why I have never had any respect for accepted ideas simply because they were accepted, or for people simply because of the positions they held."

He likes to jump into whatever area of science interests him, make his contribution, and then move on to the next area that strikes his fancy. Within the past few years, he has published studies in such diverse areas as boron chemistry, biosynthesis of fatty acids, superconductivity in oxide ceramics, phenyl radicals, and of course, recent improvements and developments in new theoretical procedures (i.e., AM1). Dewar loves science and feels that "Organic chemistry is the best preparation for anything."

Dewar is also deeply committed to establishing the validity and value of his concepts. How else can one explain his continuing, vigorous, passionate development of semiempirical methods, followed—one by one—by a series of exceptional semiempirical methods. It is fascinating to watch from the sidelines as the battle wages between Dewar (and first PNDDO, then

MINDO, then MNDO, and now AM1) and the "ab initioists". "I had never previously had trouble with referees . . ." he observes. (I can only imagine what it must be like for those who have been on the receiving end of Michael's crusades.)

I asked Dewar about errors he made over the years. "Most of my errors have been due to attempts to interpret erroneous experimental information. For example, I produced an ingenious interpretation of O_3, which accounted for it having the acute-angled structure then predicted by IR spectroscopy [in the late 1940s]. Later, an extra fundamental vibration was discovered, leading to the currently accepted structure! That [paper of mine] was incidentally (by intent) one of the worst papers ever published . . . I originally sent a short note (all it was worth) to *Nature,* giving my interpretation. The editors rejected it on the grounds that their referees did not think there was enough evidence for it. Now, at that time, Walsh was publishing long, very verbose papers—often about nothing in particular—full of references to Walsh, unpublished work; Walsh, in press; I always expected one to appear saying Walsh, personal communication. So, since anything seemed to get into print if stated at sufficient length and sufficiently padded, I decided to write a Walshian paper on ozone. *Journal of the Chemical Society* accepted it immediately!"

Dewar is a complex and brilliant man who does not by intent use his brilliance to impress others. He says what he thinks, as he thinks it, unmindful of any possible consequences. He is not afraid to disagree with people; rather, he seems to thrive on controversy. He is at once abrasively brash and charmingly boyish. He certainly does not suffer fools well, nor in silence. Some people have even called him belligerent in his stances. Even if you try, you cannot remain neutral with this man. He possesses an amazing memory and has the power to integrate his acquired and intuitive knowledge. His science is based on experimentation, but he appears to think in mathematical terms. There is no question that Dewar has made immense contributions to many areas of chemistry for 50 years.

As described in Dewar's autobiography, Michael and Mary's marriage of almost 50 years has been both a personal blessing and a professional benefit. According to his "fellow Texan", Sir Derek Barton (and friend also for almost 50 years), "His wife really understands Michael." I can personally attest to the closeness of their relationship. During the writing of this book, I often spoke with Dewar by telephone. As we wrestled with wording and expanded on memories, I frequently experienced long pauses in the conversation while the Dewars debated and concurred on a resolution to whatever matter was under discussion.

My own personal experiences with Dewar have been wonderful—full of enthusiastic vigor and cooperative exchanges. Then again, I have published a few papers in the *Journal of the American Chemical Society* and *Journal of Organic Chemistry* using MINDO/3 and MNDO to great advantage, so I am also a Dewar fan! But I suspect that even those who have, or imagine that they have, real opposition to this man also share a uniform respect. I comment on a remarkable incident that supports this belief. According to a scientist of great stature, "Upon becoming a U.S. citizen, Dewar was elected a member of the National Academy of Sciences in record time, despite having some less-than-strong personal supporters among the membership."

In the last pages of his volume, Dewar states that "Chemistry is, or should be, fun. I have always found it so, in spite of periodic battles with the 'True Believers'. . . ." There is no bland pap in Michael Dewar's life, "bland pap" being, in Jack Roberts's words, the less challenging side of science. Dewar has been described as being, simultaneously, "endearing but insufferable". I cannot help but wonder if Dewar, in his own special way, adds excitement to the lives of all who know him.

JEFFREY I. SEEMAN
Richmond, VA 23234

October 21, 1991

A Semiempirical Life

Michael J. S. Dewar

Early Years

My career in chemistry began as an indirect consequence of an early interest in science fiction.

I was born in Ahmednagar, India, on September 24, 1918, my Scottish parents being there because my father was in the Indian Civil Service, the British government of India, when it was still part of the British Empire. My early years were spent in a remote area in the Central Provinces where my father was the District Commissioner. There were only about 30 British in the place and no European children, so I had a rather strange upbringing. In those days, there was no social intercourse between the British and the people of India, not even among children, so I had no companions other than adults. I was playing bridge (auction in those days) at the club when I was 6. So when I was sent back to England to a boarding school at the age of 8, as all British children in India were at that time, it came, to say the least, as rather a shock.

British schools in those days were totally uncivilized, and I did not even have the consolation of seeing my parents for the holidays. Air travel had not, of course, been invented, and the voyage by sea took 3 weeks. There was therefore no question of children going back to India for holidays, and people in India took home leave in England only every 10 years or so, when they had saved up enough leave to make the journey worthwhile. In fact, my father died when I was 9½, being the first of many family and friends to be killed by smoking. My mother then returned to England. Supporting me through

Not yet a chemist, at 2 months with my father.

My mother.

private school would have been rough for her on her small pension, so it was up to me to help at the age of 13 by getting a scholarship to some public school (as the private high schools in Britain are quaintly called).

Privileged childhood in India. Chanda, about 1924.

Winchester College and the Classics (1931–1936)

Winchester College was (and is) one of the leading public schools in England. It was founded many centuries ago by William of Wykeham to provide poor scholars with education. The College itself still houses 70 scholars who live almost for free. Scholarships to Winchester were therefore sought after more than those at any other school, and the standards were naturally very high. Winchester also included a number of nonscholars, 400 at that time, who were distributed around 10 so-called Houses; College was the old part of the school where the scholars lived. Admission even for these nonscholars needed higher ability than admission to any other school in England. Everyone, including me, was therefore astounded when I not only got a scholarship to Winchester but came out at the top of the list. The reason for this was, in fact, simple.

While the basic object of the scholarship examination was to find the 14 best 13-year-old classical scholars in Britain, the classics (i.e., Latin and Greek) being the basic subjects in British private schools, the scholarship examination also placed emphasis on general knowledge and originality. Special weight

Everyday dress at Winchester College, not exactly a jeans-and-sneakers school.

was therefore given to an essay question designed to test the latter. The subject I was asked to write about was, "What do you think will be the most exciting new invention in the next hundred years?" Already a science fiction afficionado, I was able to come up with a simply smashing idea that I had read a few weeks earlier, and because the examiners were much too high-brow to read that sort of stuff, they thought my idea was original. I was also helped by the hours I had spent poring over the *Children's Encyclopaedia*. This was a wonderful book, in 12 huge volumes, almost as big as the *Encyclopaedia Britannica*, edited by a genius called Mees, who had a perfect understanding of children's minds. In other words, it was full of vast amounts of factual information about almost everything. Later, we reared our own children on it. Now I imagine it has been killed by television.

I cannot overemphasize my debt to the *Children's Encyclopaedia*. Thanks to it, I very early learned that the best way to find out about anything is to read about it. Whenever I asked the kind of maddening questions that all children ask, I would be told to go and find the answer myself in it. From the time I started reading fluently, before I was 5, it never occurred to me that there was any knowledge that could not be acquired by reading an appropriate book. As a result, nearly all my knowledge of science in general, and chemistry in particular, was acquired by reading. The only subjects I was really taught at school and university were ones I hated, like Latin and Greek.

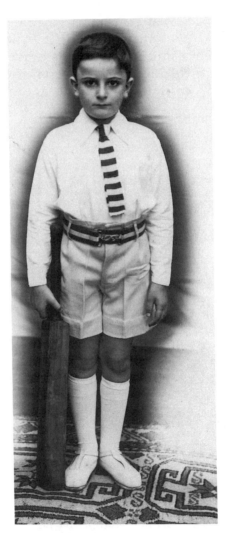

At Copthorne School, about 1928.

College at Winchester was a unique community of exceptionally intelligent people. One of the science masters, as a private project, kept records of intelligence tests of the students for many years. When he left, a friend of mine, who returned to Winchester as a science master, continued the study. Over many years, all but two of the scholars tested were off the scale of the standard intelligence test they were using. Even the two exceptions were in the genius category, near the upper limit of the test. One of them was at Winchester when I was, a year ahead of me. The rest of us all thought him rather dim witted.

Is intelligence inherited? That is a question that I have no intention of discussing here! However, it certainly seems to have been in my case. My mother had four brothers, all of

Winchester College, 1931, my first year. As the first scholar in my year, I am in the middle of the front row.

whom got B.A. degrees with first-class honors in classics at Oxford, which was then regarded as a sign of the highest intellectual distinction in England, while my father got a B.A. with first-class honors in classics at Edinburgh, the leading university in Scotland. My mother had no chance to compete because my grandmother kept her at home. One of my uncles (Berriedale Keith) also got a first in classics at Edinburgh, filling in time before going to Oxford. He won a scholarship to Oxford, to Balliol College, when he was 16 but could not go there until he was 18. After leaving Oxford, my father and the four Keith brothers all went into the Civil Service, having scored high marks in the Civil Service Examination. This was a very competitive affair because positions in the Civil Service were highly sought after. Berriedale not only headed the list of candidates but did so by an all-time record margin, scoring 30% more marks than the runner up. The other three brothers went to Burma, where one of them (William) was for a time governor. Berriedale went to the Colonial Office in London, which he left some years later to take up the Regius Chair of Sanskrit at Edinburgh University. He became the leading authority in the world on Sanskrit and also the leading authority on British Constitutional Law. I should perhaps add that Sanskrit is the ancient Indian language from which the Indo-Germanic languages were derived. These include all the European languages except Finnish and Hungarian.

Berriedale was a remarkable character. He could do two entirely different things at the same time. While he was lecturing on Sanskrit or constitutional law at Edinburgh, he used to deal with his correspondence, opening and reading letters and writing replies. Visitors to Edinburgh used to be taken to his lectures to watch. He was responsible for numerous definitive works on both Sanskrit and British constitutional law, which he typed himself, using two fingers, while his wife read him detective novels. He used to listen intently, making comments from time to time. His favorite character was Nero Wolfe. When his wife died, he hired a secretary to help him with his writing. However, since she did not know any of the technical terms, she typed more slowly than he did. So she was relegated to reading him detective novels while he typed.

Returning to my time at Winchester, life in college was a very communal affair. We slept in dormitories and shared large communal rooms, about 12 of us in each, of all years, including two prefects who were in their final year and were responsible for maintaining law and order. The prefects each had a corner of the room, while the rest occupied small open cubicles around the walls. There was little or no real privacy. Everyone was exposed to the discussions that went on much of the time about almost everything. As we all had very wide interests, the place acted as an intellectual hothouse.

One of the many things I gained there was a love of music. Only prefects were allowed to have gramophones, so the music one heard depended on their taste, which was fortunately mostly classical. Others, however, also contributed to the pool of records. So one's initial exposure to music was very much a matter of chance, and my early musical interests were a strange mixture. I remember the first lot of records I bought when I was about 16; Dvořák's New World Symphony (then Number 5, now Number 9), Rachmaninov's second piano concerto, César Franck's violin sonata, Prokofiev's third piano concerto, and Schubert's incidental music to Rosamunde. They are all still among my favorite works. If it does not sound like much of a collection, it must be remembered that records then were very expensive. A symphony or concerto occupied four or five of the old 78-rpm records and cost about $5, equivalent to $100 today. All my savings went into buying those records. The cost of records is one thing that has changed for the better.

My musical skills have mostly been confined to listening to music rather than attempting to generate it. While I had piano lessons at my preparatory school and later made an abortive further attempt to learn to play the piano, I simply lacked the necessary skills. Later I tried the violin, with even less success. My attempts to learn the violin were later terminated by an accident at Oxford. Everyone, including myself, was relieved when I stopped practicing. My only other musical activity was singing, but only in choirs. I got to know all the major choral works through singing for many years in the Oxford Harmonic Society.

Living in College was very enjoyable. Whether it was good for us is another matter. The trouble was that it undermined our self confidence. One cannot think oneself a genius in a society composed only of geniuses. Extraordinarily few Winchester scholars have made the kind of mark in the world that they should have done, given their intelligence. Most of them have aimed at finding nice, quiet, safe, intellectual backwaters. The judiciary, for example, has always been full of them. I have certainly suffered from a reluctance to publicize myself or my work. Having original ideas serves little purpose if nobody gets to know about them.

Because many of the scholars at Winchester were by no means well-off, one of the things intended to help them was a one-room second-hand book shop in the College, to which people sold school books when they left, and where the new generation could buy them cheaply. While looking for some textbooks I needed in my first year, I saw a book on chemistry and bought it out of interest. I have no idea now who wrote it. It was the standard text used at Winchester for those specializing in science. As soon as I read it, I was hooked. I still remember the excitement I felt over Avogadro's hypothesis. What a wonderfully rational way for molecules to behave! The trouble was that I soon knew everything both in it and in an accompanying introduction to organic chemistry, from cover to cover, and there were no other books on chemistry in the book shop. The school library had no books on chemistry; they were kept in a library in the science building that was reserved for those in the science Sixth Form. Fortunately one of the science masters heard of my predicament and got me a special dispensation to borrow books from the science library. I read every chemistry book in it, including two of the standard university textbooks on organic chemistry and a huge reference work on inorganic chemistry, in at least a dozen large volumes, by, if I remember right, Mellor. I am sure I am the only person who ever read it all the way through. By the time I went to Oxford, I had already decided that organic chemistry was my thing.

Any reader unfamiliar with the British educational system must be getting a bit bewildered, so I shall explain the system in British private schools as it was in my time. As far as I know, things are in fact much the same now as they were then; however, I have been largely out of touch since we moved to the United States.

The school period (ages 8–18) was spent half at a preparatory school and half at a public school, entry to the latter being by a common written examination taken at age 12 or 13. The preparatory schools provided a fairly general education with an emphasis on classics (Latin and Greek). At that time, science was not taught at all in most preparatory schools. However, I am sure things are different now.

Education for the first two years at public schools was also fairly general. However, the students were divided into several streams, each concentrating on some particular general topic, such as classics, science, history, or modern languages. For the last three years, the better students entered Sixth Form, where the specialization was much more intense. Thus those in the classics Sixth Form spent little further time on science or mathematics.

The school year consisted of three terms, each of about 3 months, with month-long vacations at Christmas and in April and 2 months in the summer.

I spent five terms in classics before switching to science. At that point, my chemical knowledge already went beyond the school level, and by the time I went to Oxford, my knowledge of organic chemistry was up to the B.A. level. As a result, I was essentially self-taught so far as chemistry is concerned. If my thoughts about chemistry have often been unorthodox, this is probably the reason. Equally, if I had not won the scholarship to Winchester, I might never have found out what a splendid subject chemistry is. Chemistry was one of the few weak points in the *Children's Encyclopaedia*.

However, I am glad I did not discover chemistry sooner. If I had, I might not have struggled with Latin as long as I did. Indeed, I think everyone should be taught Latin at school because it has three unique advantages.

First, it is a very complex language with hundreds of grammatical rules and thousands of exceptions to them. It therefore provides a superb intellectual training for the young, teaching them how to deal with clear-cut problems on the basis of pure reason. It also does not depend on specific ability, like mathematics. Anyone can cope with Latin.

Second, most children hate Latin. I certainly did. So, it teaches them at an early age to work hard at something they hate, to get to a wanted goal. Wanted goal? After once or twice having to write out 300 times *"fons* and *pons* are masculine"*, or whatever, one quickly gets the point.

Finally, Latin for most people is completely useless. This is important because one does not want to make children hostile for life to something that might be useful to them later. As a result of my early training, I have never found chemistry a burden. Any time I have to tackle some routine drudgery, I have only to say to myself "at least, it's a lot better than Latin".

Chemistry at Oxford (1936–1945)

In 1936 I went to Oxford University. I went to Oxford rather than Cambridge, and to Balliol College rather than any other, because Balliol had a scholarship in science open only to candidates from Winchester as well as their open scholarships. By winning both, I was able to, more or less, support myself. Things further improved in my second year when I got the Gibb Scholarship, the university's prize for chemistry. This had not been won previously by a second-year student, a record that, I am told, still stands.

The chemistry course at Oxford was unusual in three respects. First, it involved nothing whatsoever outside chemistry. Second, it lasted 4 years instead of 3, as elsewhere. The degree examinations were at the end of the third year. The fourth year involved only a thesis reporting original research. Third, Oxford differed from other universities in Britain in that all the teaching was carried out via tutorials. There were lectures, but attendance at them was voluntary and most of those in chemistry were attended by less able students. The tutorial system at Cambridge was much less exclusive, and at other universities it did not exist. In all my time at Oxford I only

At Oxford, 1936.

went to two lecture courses, one by Robinson and one by Sidgwick. Their lectures contained material that one couldn't get otherwise. Sidgwick was then writing his *magnum opus* on inorganic chemistry. Robinson never gave a full account of his electronic theory in print; his only review[1] appeared in a rather obscure journal. The reprint of it, which Robinson later gave me, is one of my most treasured possessions.

The lack of any instruction in mathematics was, of course, both strange and deplorable. Fortunately, the course at Winchester went to a fairly high level (partial differential equations, matrix algebra, projective geometry, etc.). Nevertheless, I have always found the lack of further formal training a serious handicap.

When I got to Balliol, I was already sold on organic chemistry. My tutor in Balliol, R. P. (Ronnie) Bell, made great efforts to convert me to physical chemistry, but it was a hopeless task. The last straw was a term spent on quantum mechanics, the only effect of which was to make me even more certain that physical chemistry had nothing to recommend it. Later, when my interests turned to physical and theoretical organic chemistry, I wished I hadn't been so inflexible.

At Balliol, the only fixed commitment was to write an essay once a week on a topic set by one's tutor and to spend an hour going through it with him. This, however, was no sinecure. I still remember my first assignment, by Ronnie Bell, an essay on the determination of atomic weights, with a reading list of 32 papers, 28 of which were in German. I had had only one short course in German at school, so it was a busy week.

Balliol was an interesting place in the 1930s. It had a long-standing reputation for academic superiority, inherited from Jowett, and it had also become a center for left-wing politics in the university. I did not myself get involved in politics. Having no preconceived ideas, I joined all three political clubs (Conservative, Liberal, and Labour) when I first went to Balliol but resigned from all three at the end of my first year in disgust. Several contemporaries of mine at Balliol later became major political figures in Britain; the most notable was Ted Heath. When I say most notable, I mean that he later became the most notable. At Balliol, he seemed to us nice but not at all impressive and not really up to the intellectual standards of the place. The general view was that he could not possibly have got into Balliol had he not won a scholarship to play the organ in chapel. The next most notable, Dennis Healey, was very different, an extremely brilliant man with wide interests whom everyone liked. I think Britain would now be a much better place if Dennis had become prime minister and Ted had languished on the back benches.

Another contemporary of mine who later achieved eminence was George Malcolm, who has for many years been one of Britain's leading musicians. I got to know George in my first term at Balliol. Because I was still trying to learn the piano and violin, I rented a piano, which George took to playing. He usually arrived at 1 a.m., when I was thinking of going to bed, and played for a couple of hours. At that time his musical interests stopped at Mozart, with few exceptions. One of these was the César Franck violin sonata, to which he had been introduced by a student of Thibaud. George used to play the piano while whistling the violin part, with a delicate vibrato. He is the only person I have ever met who could whistle really musically. Curiously enough, my Cortot–Thibaud recording (which I still think the best performance yet on records) missed many of the subtle nuances George had learned indirectly from Thibaud.

George did not know Beethoven's music at all when we first met. As I had a complete score of the Beethoven sonatas, I persuaded him to have a look at them. Because the first sonata (Op. 2, No. 1) is very like Mozart, George was easily hooked. Over a period of weeks, he played the whole lot, sight reading them and making very few errors. It was a spectacular tour de force and very enjoyable because they were really good performances. I still remember the night he played Op. 106 (the Hammerklavier). By the end, George agreed that Beethoven had quite a lot to offer.

The sudden freedom from school regimentation allowed us to develop all kinds of interests and explore all kinds of activities that had been effectively barred at school. My main new one was bridge. I had learned auction bridge, the precursor of contract, in India when I was 6, and when my father died and my mother came to England, when I was 9, I used to play first auction bridge, and then contract, with her friends at the hotel in London where we stayed for several years until she got a flat. When I went to Winchester, I met some other young bridge players who lived in London, and we used to play together in the holidays. However, I had never taken to the game really seriously, apart from reading Culbertson's books, until I went to Oxford. One of my Winchester friends, Edmund Phillips, was not only also a chemist but also went to Balliol at the same time that I did. We found that there were only four other bridge players in Balliol and they only played family-style bridge once a week. Edmund and I soon changed the situation. By the end of our second year, a quarter of the students at Balliol were playing bridge seriously nearly every night. The Master of Balliol once remarked that no two people in the history of the College had been responsible for so many people getting Seconds and Thirds in their degrees when they should have got Firsts! In our second year, we founded a University Bridge Club and played a match against Cambridge. Because our team was completely inexperienced, it was not surprising that we lost by a rather large margin. The following year we were much better organized. We lost again, but it was very close; the match was decided on the last of 100 boards. At this point, World War II intervened, so we did not play Cambridge again. Students within 1 year of graduating were allowed to stay on to take their degrees, but after that the university effectively closed for the duration. We did play a friendly match against the English International Team in the winter of 1939 and won by a large margin, so I think we might have beaten Cambridge in 1940 if we had had the chance.

At this point everyone was drafted. I was told to stay on at Oxford to carry out war research, but all my bridge-playing friends left and our bridge club dissolved. The only way I could play bridge after that was at a club for stakes that were too high for me and, besides, I have always hated playing games for money.

From then on I played bridge only at occasional tournaments, with Edmund, and we had a number of successes. Tournaments in Britain were, however, held in expensive hotels, so when my sons went off to school and my wife and I were faced with crushing school fees, I gave that up too. Since then I have confined my bridge to reading.

Instead, I took to chess. Oxford was a good place for chess during the war. The Oxford City Chess Club had been good to start with and it had been further strengthened by an influx of refugees from Europe who had been leading chess players. Chemistry and chess seem to go well together. Both Kappa Cornforth and Sir Robert Robinson were enthusiastic chess players. Indeed, Kappa held—and I think still holds—the amateur world record for the number of games played simultaneously blindfolded. I never myself reached any great heights, but three of my games in Oxford County Championships were published in *The British Chess Magazine*, one a titanic struggle

With Kappa Cornforth.

with Robert Robinson that ended in a draw (he usually beat me) and one with a new move that refuted the standard line in a not-very-popular opening. My excursion into chess ended when we left Oxford because there was no chess club in Maidenhead. I have never played serious chess since.

By my fourth year, I had become totally absorbed in natural-product chemistry, in particular in alkaloids, then the area of most topical interest. I therefore decided to carry out my fourth-year research on the structure of yohimbine with F. E. King, who had been my organic chemistry tutor. At this point, however, Hitler was inconsiderate enough to start World War II. Everyone in Britain was immediately drafted, and I was ordered to stay in Oxford to carry out research on war problems. Yohimbine clearly did not qualify. Thanks to the assistance of some friends, I was, however, able to make one small contribution concerning it.

Yohimbine has been claimed to be the active agent in *Quebracho yohimbi*, a West African plant reportedly used by the natives as an aphrodisiac. As I had obtained an ounce of yohimbine hydrochloride that I could not now use in my planned research, six of us, all fourth-year chemists (including a very attractive girl), decided the matter should be properly investigated. We, therefore, organized a yohimbine party. Unfortunately, the results were totally negative. Although we ended by taking three times the recommended dose, we had to conclude that if *Quebracho yohimbi* is in fact an aphrodisiac, yohimbine is *not* responsible.

All I can say about my contribution to the war effort is that it did not seriously impede the Allied cause, and the work did get me a D.Phil., as Oxford whimsically terms its equivalent of a Ph.D. My first project was to try to find new and better explosives. The people at Woolwich Arsenal, the center of explosives research in England, were, however outraged at what they regarded as outsiders trespassing on their preserve. The fact that a war happened to be in progress seemed to them no excuse. They solved the problem very simply, by failing to test any of the compounds I sent them. I still think one of them (1) might be interesting as an explosive; unpublished calculations by Jack Alster at Piccatinny Arsenal, in New Jersey, support this idea.

Next, I spent some time trying to make new sulfa drugs. Sulfapyridine (2) had been discovered just in time to prevent thousands of deaths from flu-induced pneumonia in the first year of the war, and the hunt was on for better analogues. I, therefore, made the corresponding derivatives of 3- and 4-aminopyrazole.[2] Both, however, turned out to be too insoluble to be tested.

1

2

The Dyson Perrins Laboratory

At this point, having got my D.Phil., Robinson arranged a post-doctoral appointment for me to work with him on further war problems. He had, in fact, been annoyed at my not choosing to work with him earlier, for my B.A. and D.Phil. However, as I told him, it would have been hopeless for me to work with him at that stage, because what I needed was experience at experimental work with detailed supervision that he could not have provided because he was hardly ever in Oxford. During the war, he came into the Dyson Perrins laboratory, or the D.P. as it is commonly called, roughly once a month. He then got through as much work in a day as any normal person would in a month, so it worked out well, provided one needed only advice and help on a very high level.

I always had a great affection for Robinson and an enormous respect for him as a chemist. He really was incredibly gifted. He could see in a flash solutions of problems that had totally baffled very good organic chemists. It is a tragedy that he never wrote a single book about chemistry. If he had presented his electronic theory in book form in 1930, as, of course, he should have done, the history of organic chemistry would have been very different. I will tell just one of the many stories about him.

On the one day a month that Robinson was in Oxford, he was naturally besieged by his research group. While he talked to them about their problems, he used to deal with the correspondence that had collected while he was away. On one occasion, he was talking to Arthur Birch about an abstruse problem in steroid chemistry, when he suddenly burst into laughter at a letter he was reading. This letter was from an American chemist who had been trying for 6 months to deduce the structure of a degradation product from some alkaloid and had been completely unable to come up with a reasonable structure for it. It may seem strange nowadays that such a thing could happen; however, life was much more complicated then, when spectroscopy was unheard of and structures had to be determined entirely from elemental analyses and chemical reactions. Anyway, Robinson solved the problem in a flash. As a postscript, Arthur later tried it on everyone in the D.P. and no one came up with the solution. (If you would like to have a go, *see* Appendix A.)

At that time there was a major drive in progress to find new antimalarials for use in the war in the Far East. I was given the assignment of trying out new heteroaromatic rings. I made

I was originally a respectable experimental organic chemist. Here, at Dyson Perrins Laboratory during World War II.

a large number of such compounds with aminoalkyl side chains. Unfortunately, not one of them showed any activity of any kind against any known microorganism, let alone against malaria. This work was published[3] at Robinson's behest, a friendly (to me) action that involved him in angry recriminations because it had been agreed that none of the antimalarial work in Britain or America should be published until the end of the war. Presumably, his excuse was that none of my compounds could reasonably be classified as antimalarials.

Penicillin had now become the top priority, and I was transferred to the penicillin team. I was given the task of making an analogue of penicillin with the saturated thiazolidine ring replaced by thiazole. Unfortunately, the then-accepted (oxazalone) structure for penicillin was incorrect, so even if I had succeeded in making the compound I was trying to make, which

Sir Robert Robinson, one of the greatest organic chemists of all time. I owe him much. He was also an enthusiastic chess player. One of my many happy memories is an epic match Robert and I played one year in the Oxford County Championship, a titanic struggle that ended in a draw after more than 80 moves. It was published in The British Chess Magazine.

I didn't, it would, in fact, have been an analogue of something else. One incidental observation of some interest came out of this work. Alkaline hydrolysis of 3 gave a salt of the corresponding acid that was stable in solution. Acidification, however, led to immediate decarboxylation at room temperature. Presumably, the reaction involved a phenomenally facile retroene reaction (cf. 4).

At this point came the most important event of my life. I was lucky enough to meet someone who was more than my equal, Mary Williamson, one of only three people in her year to get a B.A. in history at Oxford with First Class Honours. We had seen each other around in Oxford, having mutual acquaintances, but we met only after she had left, and then by chance. She had returned to Oxford with her family for the degree ceremony and called at my lodgings in the hope of seeing a boyfriend who was still at Oxford and also lived there. Because I happened to come in just as she found he was out, she asked me to lend her a pencil to write him a note. In return, I invited her in for a drink, and 18 months later we were married. Although Mary's career in history was sabotaged by the war and by marrying me, she is now well recognized in her field (English Tudor history), and her books are widely quoted. The fact that she is a historian and not a scientist has been of benefit to both of us because we both have very wide and intermeshing interests, because we tend to look at things from different viewpoints, and because having both been reared in the Oxford tradition of medieval scholasticism, we have both always treated argument as a way to find whether our ideas are right rather than as a form of personal confrontation. Because this now seems to be an unfamiliar attitude, particularly in America, perhaps I should say more.

Wartime wedding, June 3, 1944—three days before D-Day.

Medieval scholars were taught the art of argument by set confrontations in which two protagonists were assigned to take opposite sides, irrespective of their own feelings about the issue in question. The subjects chosen were often frivolous (for example, "How many angels can dance on the head of a pin?"). The object of these exercises was to make sure that arguments about matters of real substance would not be swayed by incompetence on the part of the participants due to their inadequate training in the art of argument. The best way to test an idea or point of view is to defend it against all comers. If, in spite of one's best efforts, one is unable to defend a position, then one must abandon it, even if it involves a cherished belief. Equally, to be effective, verbal battles of this kind must be conducted ruthlessly and with no holds barred.

In our case, almost any statement by one of us has always tended to be contradicted by the other, starting an argument that continues, often ferociously, until we reach an agreed solution. As a result, neither of us has ever been able to get stuck in mental grooves. With the passage of time and innumerable arguments about everything under the sun, we have naturally come to think the same way about most things. However, this does not mean that our ideas have become fixed because our arguments still continue. We are just as happy to defend some new viewpoint to see if it is right as to defend our current beliefs. Even now, when we are both over 70, we still find ourselves changing ideas we have held for decades in response to new evidence or new thinking.

As an aside, I think this should always be the approach to argument among civilized and intelligent people. Unfortunately, it rarely is. Most people tend to regard any contradiction of their own beliefs as a kind of quasi-religious confrontation, and any change of basic beliefs on anybody's part as heresy or plain treachery. This is certainly true in chemistry! I am one of the few chemists who has changed my basic chemical beliefs in response to new evidence or new arguments, and I have done it repeatedly. Equally, the fact that I argue ferociously about chemistry, with no holds barred has upset many chemists. To me, it is the only way to find out whether or not a given theory or idea is right. If one puts one's best efforts into defending a pet idea and yet fails, then one has to abandon it.

One of my early benefits from meeting Mary was being introduced to Collingwood. I had become interested early in philosophy and had read all the standard works. After struggling with Kant's *Critique of Pure Reason* and Whitehead's *Process and Reality*, surely two of the most unnecessarily obscure books ever written, Collingwood came like a breath of fresh air. His *Metaphysics*, *The Idea of History*, *The New Leviathan*, and *The Idea of Nature* not only made perfect sense to me, but also for the first time made sense of what I had already read. He seemed to me the first philosopher to really understand what science is, or at least should be, about. I know this is an unfashionable view. All I can say is that my own work has profited greatly from the principles I learned from him. To describe these in detail would take a book in itself. Here I will give just one example, concerning the nature of scientific "knowledge".

As Collingwood points out, the whole edifice of science rests on a set of basic assumptions ("absolute presuppositions") that cannot be proved true or false. They are simply assumed. Indeed, they change from time to time. Thus, current ideas in science concerning causality differ from those of Aristotle. In science today, it is assumed that an event is caused only by events in the past, that is, by *efficient causes* in Aristotle's terminology. According to this view, the world is propelled forward from behind, the future being determined by the past. However, Aristotle also considered that events might be caused by potential events in the future, the driving force being a motive to reach some desired goal. Such final causes are of course well recognized in everyday life. People's actions are frequently guided by the desire to reach some end rather than by prodding from behind. If you ask a scientist why science does not allow

Confrontation? With Arthur J. Birch at the Ciba Symposium in honor of Sir Robert Robinson, 1977. Arthur was another member of the Dyson Perrins gang during World War II, in the lab next door to me and the Cornforths. Rear center, left to right, Jack Baldwin, R. B. Woodward, and Lord Todd.

for the possibility of final causes operating, he or she will reply, "Everyone knows that they don't." If you press him or her for evidence that this is the case, the result will be either a lot of bogus arguments and irrelevant facts, or anger and abuse because there is no such evidence. Since the present definition of causality is one of the absolute presuppositions of current science, its truth is not open to discussion.

Because the basic assumptions on which the whole edifice of current science rests are neither true nor false, the same must of course also be true of science itself. Why then do absolute presuppositions change? Collingwood stops short of discussing this, but the answer is fairly obvious. The stated object of science is to explain the universe. Because science, by its nature, is precluded from being a search for ultimate truth, any such explanation must be stated in the form of a model. A model is a simple mechanism that simulates the behavior of a more complex one. A scientific model must simulate the behavior of the universe, or some part of it, while remaining simple enough for us to understand. The test of such a model is purely operational. Does it in fact simulate the behavior of the system being modeled? If not, we have to modify it or replace it with a better one, better in the sense that it simulates the parent system more effectively. There is of course no question of a model being true or false. The same rule applies to scientific theories, which are simply definitions of scientific models. The question "Is it true?" is meaningless in science. The correct question to ask is "Does it work?"

Aside from That . . .

Many scientists today treat science as a kind of religion, regarding any criticism of their scientific beliefs with the same sense of outrage as a good Catholic would greet criticism of the dogma of the virgin birth. Instead of trying to move science forward by investigating new alternative models (i.e., theories), they treat dissenters with the kind of zeal reserved by most churches for unbelievers and heretics.

A good example is provided by the way that the California astronomers treated Arp. Arp had a good reputation for his observational work in astronomy and was for many years one of the major users of the large telescopes in California. However, some years ago, he came up with photographs of distant spiral nebulae that seemed to run contrary to existing ideas concerning the structure of the universe. It is generally believed that the universe is expanding, so distant nebulae are receding at a rate proportional to their distance from us. This movement is indicated by their spectra, in which the lines are correspondingly displaced to the red by the Doppler effect. Arp found several cases where pairs of nebulae with different red shifts appeared to be linked together, and he pointed out that if this were really the case, the red shifts would have to be due to something other than the Doppler effect. This conclusion outraged other astronomers, to whom the official picture of the universe had become a basic article of faith and one on which their careers depended. Although nobody has been able to give a convincing alternative explanation for Arp's photos, the astronomers in California hounded poor Arp, and when (like Galileo!) he failed to recant, they succeeded in getting him banned from using their telescopes, thus virtually terminating his career as an observational astronomer. It is very difficult to see any basic difference between the way the Catholic church treated heretics in the Middle Ages and the way the California astronomers treated Arp. It is true that they did not try to burn Arp at the stake, but that is only because times have changed.

To me, this kind of intolerance has always been utterly disgusting. It has moreover become much worse in recent years, especially in the United States, with the increasing compartmentalization of science. Each branch of science is now divided into dozens of subbranches, each with its set of members and shibboleths, and the members of each branch bitterly resent intrusions by outsiders, particularly if they involve a revision of official beliefs in the area. This situation is unfortunate because new ideas often have to come from outside, particularly in the less important or less fashionable subdisciplines that fail to attract able people. I have suffered more than most from this kind of hostility because I have an unusually wide range of interests, because I do often come up with new ideas, and because I have never respected authority for its own sake.

One of my early experiences will show how times have changed for those who challenge the basic assumptions. When I was 11, I bought, with money I had been given for my birthday, a wonderful book called *The Splendour of the Heavens,* an oversize volume with about a thousand pages and many hundreds of photographs that gave a remarkably complete summary by leading astronomers of astronomy at the time it was published (1925). This book began an interest in astronomy that I have kept ever since. At Oxford, I read most of the more recent books on astronomy in the Radcliffe Library (the university's science library) as well as most of the British and American journals on astronomy and astrophysics. During the war, I had an idea for an alternative to general relativity theory, based on a combination of special relativity theory with the assumption that the velocity of light effectively varies, being fixed relative to a local frame of reference at each point. Looking through the literature, I found a forgotten paper that showed that all the classic predictions of general relativity theory can in fact be explained in this way; I also found direct evidence for my suggestion in the form of an apparently anomalous regularity that had been reported in the orbits of spectroscopic binaries. I mentioned this one day to Sir David Ross, who was the vice-chancellor of Oxford and a leading Aristotelian philosopher, and also one of my mother's cousins. Sir David was interested, never having felt that Aristotle and general relativity mixed, but he could not of course assess my suggestion. So he arranged for me to meet Milne, the professor of astronomy at Oxford, a leading theoretician whose interests lay primarily in cosmology and relativity theory. I spent an afternoon with Milne, explaining my wild ideas. As time went on and I was able to meet all his numerous objections, he became more and more agitated. Finally he said, "This is clearly nonsense. Anyone can see that it is nonsense. But I can't see why it is nonsense. You must publish it so that other people can find what is wrong with it." So a paper with the title "An Interpretation of Light and Its Bearing on Cosmology" duly appeared, with Milne's help, in *The Philosophical Magazine* in 1947. I doubt if that could happen anywhere today, certainly not in any American journal. The referees would recommend rejection, even if they were unable to find any valid criticism, and the editor would accept their recommendation.

Had my idea any merit? That is difficult to say. The original version was virtually ignored. I had just one request for a reprint. It was also rather naive. Since then I have improved and elaborated it, and I think my present version has a lot going for it. However, I have no intention of trying to publish it.

Astronomy is one of a number of interests that I learned about by reading the *Children's Encyclopaedia* and later followed to a high level. Another, geology, was stimulated later by an accident in the laboratory at Oxford, during World War II, an episode that also has a lesson to tell.

The accident was due to sheer carelessness. I stuck a piece of glass tubing through my left hand and severed both flexor tendons of my left forefinger. I am sure that nowadays developments in microsurgery have made such injuries trivial. At that time, however, damage to flexor tendons (the ones on the palm side of one's hand) was a disaster. Fortunately, the surgeon who dealt with me was a leading expert on tendon injuries and was trying a new way of treating them. At the end of 2 weeks, I had recovered complete movement in the finger. If I had stayed that way, I would have been unique in the medical literature. At this point, however, the stitches dissolved and the tendons came apart. Later I realized that I had all the symptoms of incipient scurvy, due to eating a shockingly poor diet. Lack of vitamin C prevents wounds from healing. As I was working on chemotherapy at the time and knew all the relevant literature, eating so poorly was sheer stupidity on my part. Anyway, after two further operations and thanks to another new experimental surgical procedure, I did recover partial use of the finger. The only major consequence was that I had to give up trying to learn to play the violin, which was really a blessing to everyone, including myself.

One lesson of course is that one should take massive doses of vitamin C before undergoing surgery. Because of the determination with which doctors and surgeons have rejected claims concerning the beneficial effects of vitamin C, hospitals still fail to give patients adequate doses. My sister not long ago had an accident in which she broke a leg in London at the age of 85. After 2 months in one of the leading London hospitals, the injury was still not healing, and her surgeon told her that there was little hope that it would. As she has always been very active, this would have been a disaster. However, at this point she got a friend to smuggle vitamin C to her and started taking massive doses. In a few weeks, her bones had joined up nicely, and she was soon back to normal again. Her surgeon refused to believe that vitamin C had been responsible.

If vitamin C were an expensive prescription drug, it would have been hailed long ago as one of the greatest medical discoveries in the history of humankind. There is no question concerning its efficacy for treating colds and flu. The so-called "tests" that have been carried out by the AMA have been designed not to test it but to discredit it, using wholly inadequate dosage and refusing participation by anyone who did not oppose it a priori. I have gained greatly by introducing it to my research group and to my secretary in the form of greater efficiency and fewer days lost through sickness. My own experience was sufficient to prove this many years ago.

I might add that the effects of vitamin C were not discovered by Linus Pauling, though he is clearly a good recommendation for it. The effects were discovered in Britain by organic chemists during World War II. Early on, the word spread that taking a few hundred milligrams of vitamin C before going to a party stopped one from having a hangover (which is true, but a gram is better), and later this was extended by the news that taking a gram of vitamin C when one felt a cold coming on often aborted it, which also turned out to be true. Later, when we moved to Chicago and started catching terrible colds whenever we returned to Europe, having lost our resistance to European germs, it occurred to me that taking really large doses of vitamin C might meet the situation—and it did. Try it! If you feel nervous, there is an article in a highly respectable ACS journal (*CHEMTECH*) that you may find interesting: an interview with Dr. Cathcart, in the February issue of 1978.

Anyway, I was unable to work in the laboratory for more than 3 months, so I had time to spare after reading my usual quota of chemical journals. I spent the time reading all the more recent books about geology in the Radcliffe Science Library, and a book by Brooker led me to a novel idea about the way mountains are formed.

At that time, few people believed in continental drift. It was thought that mountains were formed by contraction of the earth as it cools, like the wrinkles on a drying apple. However, several lines of evidence seemed to indicate that the wrinkling is not continuous. Major episodes of mountain formation seemed to occur periodically, roughly every 200 million years, and between these periods the earth seemed to expand. However, nobody had been able to devise a possible mechanism for this alternate expansion and contraction. The best candidate was one proposed by Joly, involving periodic melting and resolidification of the layer of basalt that was known to lie under the earth's crust. This mechanism had, however, been refuted by the realization that such a layer would be stirred up by convection currents and could not therefore exist in a partly molten state.

It occurred to me that this objection could be avoided if the phase change involved two solids instead of a solid and a liquid. Indeed, eclogite was known to be a denser, high-pressure form of basalt, formed from basalt at high pressures. I won't go into details. It was just a simple exercise in thermodynamics. In the expansion phase, eclogite at the bottom of the layer is converted to basalt. The resulting situation is in principle unstable but persists because convection currents cannot (it was assumed) be set up in crystalline rocks. At some point, however, the layer of basalt that has been formed begins to melt. The layer then inverts, the hot, light basalt at the bottom rising and the cooler, denser eclogite above sinking. At that time, no thermochemical data were available for the basalt—eclogite system. However, making reasonable assumptions and using accepted estimates of the thickness of the relevant layers and the rate of flow of heat from the earth's interior, I was able to show that one could obtain an oscillation of the earth's radius of the magnitude estimated by Brooker, with about the right period.

Some years later, when I went to Queen Mary College, University of London, I wrote this up and submitted it to one of the geological journals (I forget now which one). I got it back with a nice note from the editor, saying that he thought it very interesting but that it was really geophysics, not geology. So I next sent it to a geophysical journal, whose editor returned it, saying that while it was very interesting, it was really geochemistry, not geophysics. My third attempt was to *Geochimica et Cosmochimica Acta*. This time the editor, after again expressing interest, said that it was really geology, not geochemistry. At this point, faced with an infinite regress, I gave up. Many years later, when I showed my manuscript to a geological colleague at University of Texas at Austin, he was quite intrigued, remarking that I must have been one of the first to suggest that a phase transition might be the driving force in crustal movements.

Now, of course, continental drift is an established concept, and it is generally assumed that the process is continuous, driven by convection currents in the subcrustal layers. This model rests on the unproved but universally accepted assumption that crystalline rocks will flow under any stress, however small, if the stress is applied for a sufficiently long period of time. I still wonder if this assumption is in fact true. I also suspect that the evidence, that major worldwide episodes of mountain formation occur at long intervals, has been brushed under the rug because it does not fit in with current ideas. In many cases, a bitter confrontation between two rival theories ends in one being routed; the winners take this result as proof that they were wholly right and the losers wholly wrong. However, it usually turns out later that the truth lay somewhere between the two extremes. Here it seems quite possible that a thermal cycle of the kind I suggested does operate, but in a deeper subcrustal layer (A). Each inversion of A leads to a major convulsion followed by a long period of cooling, during which convection currents in the basalt layer, enabled by molten rock from A, lead to continental drift.

The D.P. during the war was a very exciting place. I shared a laboratory with Kappa and Rita Cornforth, and others around included Helen Muir, Arthur Birch, and Richard Martin. Kappa, who won the Nobel Prize in chemistry some years ago for his work on the biosynthesis of cholesterol, recently retired from a Royal Society chair at the University of Sussex; Helen is now the director of the Division of Biochemistry at the Kennedy Institute of Rheumatology; Arthur returned to Australia after the war to a chair of chemistry at the Australian National University; and Richard went to Belgium after the war, to a chair at the Free University of Brussels. Kappa, Helen, and Arthur have all been fellows of the Royal Society for many years. Quite a distinguished set of colleagues! If I have not included Rita, this is only because she was a victim of the times, like Mary. Robinson thought her one of the best members of his group. However, she put marriage before her career. She has always worked with

Sir John and Lady Cornforth. I shared a laboratory with Kappa and Rita at the Dyson Perrins Laboratory during World War II. We have been close friends ever since. For his work on the biogenesis of steroids, Kappa later shared the Nobel Prize with Vladimir Prelog.

Kappa in the laboratory, and her help has been the more vital to him in view of his deafness. The problem of clashing careers in cases of this kind is one for which there is still no real solution.

While we were all working on different projects supposedly connected with the war, the Cornforths and I collaborated on one private project. A group of Indian chemists reported a synthesis of santonin, in which one step was claimed to have led to optically active material from inactive reactants. The reaction involved C-methylation of a derivative of 2-formylcyclohexanone. The authors also claimed that methylation of 2-formylcyclohexanone had given an optically active product. We repeated the reaction, and the product was, of course, totally inactive. We submitted a note reporting our results to *Nature*,[4] congratulating the original authors on achieving a result that would be expected only once in $10^{10^{20}}$ trials. The editor, however, was clearly disturbed by our wording and altered it. The final form read: "This result violates no fundamental physical laws and would indeed be expected to occur once in $(10^{10})^{20}$ trials." I don't know if this proved any less offensive to the Indian authors than the original version would have done; however, I think the change of our estimated odds by 10^{18} powers of 10 must be something of a record for editorial revision.

While I did not contribute significantly to the penicillin project, my joining Robinson's group led indirectly to my first real contribution to chemistry. Organic chemists from America, also working on penicillin, used to visit the D.P., and Robinson held joint meetings for them with his team. At one of these, before Dorothy Hodgkin's crystallographic work established the β-lactam structure, a visiting chemist (I forget who) remarked that penicillin seemed, like stipitatic acid, to be a molecule for which one could write no rational structure. Raistrick, the leading authority on mold metabolites, had worked for 2 years on the structure of stipitatic acid without being able to come up with a reasonable structure for it. The matter had become something of a *cause célèbre*, all the leading organic chemists in the West having been equally baffled. At the end of the meeting, I went to the library and read Raistrick's papers. In three-quarters of an hour I had the solution.[5]

Since then, I have come up with a number of similar contributions that have given me an undeserved reputation for original thinking. I am not, in fact, original in the proper sense of the term. I am, however, good at solving problems, for two reasons: I have a talent for asking the right questions, and I suffer less than most people from getting stuck in mental grooves. Why do I have this talent? Why does anyone have any talent?

In the case of stipitatic acid, the right question should have been easy to find. Stipitatic acid is a dibasic acid that is converted by alkali fusion to an isomeric dibasic acid, 5-hydroxyisophthalic acid (5). Clearly, one of the acid functions in stipitatic acid must correspond to a group other than carboxyl, which is converted to carboxyl by alkali. The crucial question was: What group is converted to carboxyl by alkali? Obviously, an α-diketone, so the next step was to write the corresponding precursor 6 of 5. Because stipitatic acid is a dibasic acid, the COCO group in it must be enolized. Now comes a second obvious question: The chemical evidence shows that stipitatic acid is aromatic; why should this be true of 7? In terms of resonance theory, this question can be rephrased: Can two or more resonance structures be written for 7? Clearly they can, if the ketone group is also enolized, because stipitatic acid is a hybrid of the bis-enolic structures, 7 and 8. I termed the parent aromatic system 9 *tropolone*, which is short for tropylenolone (tropylene being 1,3,5-cycloheptatriene).[5]

Stipitatic acid was the first compound to be recognized as aromatic in which the relevant ring had other than five or six members. This interpretation also led to the solution of another long-standing problem, the structure of colchicine. My familiarity with alkaloid chemistry led me to recognize the parallel between colchicine and stipitatic acid and hence to suggest[6] that colchicine, too, is a tropolone derivative, as was later shown to be the case.

5

6

My formulation of tropolone was naive by present stand-ards. Tropolone is now recognized as a derivative of the Hückel $C_7H_7^+$ polyene cation, tropylium, about which I will have more to say presently. At that time, organic chemists, including me, knew little or nothing about molecular orbital (MO) theory in general and Hückel's contributions in particular. Indeed, even resonance theory was viewed askance in the D.P. However, because the war kept me away from experimental work other than my war problems, I became increasingly interested in theoretical organic chemistry as an outlet. My resulting explora-tions of MO theory led to a second major contribution.

The Benzidine Rearrangement

One of the most baffling organic reactions was, and still is, the benzidine rearrangement (i.e., the acid-catalyzed conversion of hydrazobenzene (10) and its derivatives to benzidines (11) and other analogous products). Experimental evidence had clearly indicated these reactions to be intramolecular and extremely facile, that of 10 being almost instantaneous in strong acids at room temperature. How can the *para* carbon atoms begin to bond to one another before the NN bond has broken, given that the carbon atoms are initially more than 6 Å apart and given that benzene rings cannot approach closer than about 3.5 Å? Other products are also formed in some of these reactions (e.g., diphenylines (12) and ortho-(13) and para-(14) semidines). What determines the products from a given derivative of 10? After poring through the literature, I was able to formulate[7] a series of empirical rules that correctly predicted the products formed, and my knowledge of MO theory, while still embryonic, was nevertheless sufficient to suggest a seemingly attractive mechanism.[7,8]

There is no basic difference between an atomic orbital (AO) and an MO. Both represent, as it were, a volume of orbital space able to hold two electrons. Because AOs of different atoms can interact with one another to form MOs, MOs should likewise interact with other MOs to form larger MOs. If the interacting orbitals contain just two electrons between them, the interaction should, in each case, lead to the formation of a bond: a normal covalent bond if each contributing orbital is singly occupied, and a dative covalent bond if one is filled and the other empty. In other words, MOs should serve as effective substitutes for AOs in forming covalent bonds. The requirements are the same as for AOs, that is, the orbitals involved must contain just two electrons between them and they must overlap effectively with one another. The latter requirement eliminates most σ bonds from consideration in this connection because σ MOs are tucked away between pairs of atoms and cannot therefore, as a rule, overlap effectively with AOs or MOs of other atoms or molecules. This difficulty does not, however, arise with π MOs.

12

13

14

It was thought at the time that the benzidine rearrangement involved monoprotonated **10** (i.e., **15**). Heterolysis of the NN bond in **15** would lead to aniline (**16**) and a cation **17** derived from aniline by loss of a proton and two electrons. The highest occupied MO (HOMO) of **16** is thus empty in **17**, constituting its lowest unoccupied MO (LUMO), and both these MOs are of similar shape. They should therefore be able to interact with one another to form a dative covalent bond, leading to a sandwich structure **18** in which the *para* carbon atoms are relatively close together. Conversion of **15** to **18** should require little energy because the number of covalent bonds in them is the same. The only difference is that one of the bonds in **18** is formed by interaction of two MOs instead of two AOs. Conversion of **15** to **18** involves only a folding up of **15** and spreading of the two electrons originally localized in a two-center N-N bond into an analogous multicenter bond formed by interaction of two MOs.

Unfolding of **18** can likewise lead directly to **19**, in which the many-center covalent bond has shrunk back to a two-center C–C bond, and **19** can undergo prototropic shifts leading to **11**. The other products could be formed from analogous sandwich intermediates derived from **18** by rotation of one aniline moiety about the line joining the centers of the rings. I termed species of this kind, in which one or both of the AOs in a normal covalent bond are replaced by MOs, π complexes.

Later, it was shown that it is in fact the diprotonated form **20** of **15** that rearranges. However, this modification involves no major change in the π-complex mechanism. Homolysis of the NN bond in **20** leads to two radical cations **21** with identical singly occupied π MOs that can be used to form a mul-

$$H_2N^+=\text{[ring]}\overset{H}{\underset{H}{\text{C}}}\text{[ring]}=NH$$

19

$$\text{[ring]}-\overset{+}{N}H_2\overset{+}{N}H_2-\text{[ring]}$$

20

$$\text{[ring]}-NH_2 \quad +\bullet$$

21

ticenter covalent bond. I termed the multicenter bonds in π complexes μ *bonds*. The term μ bond was later adopted by organic chemists in a more general sense that fortunately includes mine as a special case.

The mechanism of the benzidine rearrangement remained uncertain for many years. While Shine reported studies of kinetic isotope effects which, he claimed, refuted the π complex mechanism, his new evidence could in fact still be explained in terms of it without undue strain. Later I will describe some recent work that seems to have at last solved this classic problem.

π-Complex Theory

The main significance of this work lay, however, in the basic idea. One immediate consequence of this idea was a simple explanation[8] of the ease with which carbenium ions rearrange (Wagner–Meerwein rearrangement). Such an ion (**22**) can be formally dissociated to an olefin and a cation (R^+), that can in turn combine to form a π complex (**23**) in which the filled π MO of the olefin acts as the donor and the empty AO of R^+ as the acceptor. Because **23** contains the same number of covalent bonds as **22**, the difference in energy between them should be small. Thus, rearrangement of **22** to **23**, and hence to an isomeric carbenium ion **24**, should require very little energy.

This interpretation also explains why analogous radicals and anions rearrange with difficulty, if at all. The relevant orbi-

$$R{-}C{-}\overset{+}{C} \qquad \overset{R\,\uparrow^{+}}{C{=}C} \qquad C{-}\overset{+}{C}{-}R$$

22   **23**   **24**

$$\overset{R}{C{-}\overset{+}{C}} \longleftrightarrow \overset{R}{C{-}\overset{+}{C}} \longleftrightarrow \overset{R^{+}}{C{=}C}$$

25

tals in the analogous intermediates (e.g., **23**) would contain too many (three or four) electrons. The extra electron, or electrons, would be forced into an antibonding MO (*cf.* the interaction between He and H, or between two helium atoms). This difference could not be explained in terms of resonance theory because exactly similar resonance structures (**25**) can be written for all three systems. I suggested that μ bonds in π complexes such as **23** should be represented, as indicated, by the usual arrow symbol used to depict normal two-center dative bonds.

Fired with enthusiasm by my recent explorations of Hückel molecular orbital (HMO) theory, I presented calculations[9] for simple π complexes at a Faraday discussion in 1947. Charles Coulson sent me a manuscript of a paper ferociously attacking my, admittedly, rather naive contribution. I succeeded in convincing him that my calculations were not in fact as bad as he had thought, and we ended by publishing a joint paper[10] pointing out the inadequacy of HMO theory in any connection involving ions. Although I took this to heart, others, unfortunately, did not, and a number of correspondingly meaningless calculations for ions appeared in the literature before HMO theory was finally abandoned.

If π complexes are similar to the isomeric carbenium ions in energy, cases might be expected where the π complex is, in fact, the most stable form of the ion. Indeed, Wilson[11] had suggested in 1939 that ionization of camphene hydrochloride leads to such a species. His suggestion was, however, ignored because

C. A. Coulson. Charles pioneered the use of MO theory in organic chemistry. He and Christopher Longuet-Higgins were the first to recognize the value of perturbation theory in this connection. My PMO theory was based on their work.

he was forced to formulate it in terms of resonance theory (*cf.* **25**). The idea that this π complex could be more stable than a classical isomer seemed at the time so wildly improbable that Wilson had difficulty in publishing his suggestion, even tentatively in the form of a single sentence! I interpreted the intermediate as a π complex in my first book,[12] and the following year, at a conference at Montpellier in France, I presented a general review of π complex theory[13] in which I elaborated this point.

At the same meeting, Saul Winstein presented his evidence for the "nonclassical" nature of the 2-norbornyl cation, which he clearly regarded as a major contribution to chemical theory. He was obviously put out by my claim that this was no more than a confirmation of ideas that I had already presented in print. He never accepted my representation of such species as π complexes or acknowledged my contributions. Indeed, when he later published a paper with Simonetta,[14] reporting HMO calculations for a "nonclassical" carbonium ion, as carbocations were then called, he failed to refer not only to my earlier calculations[9] but also to the accompanying paper by Coulson and me,[10] in which we pointed out the inadequacy of the approach they had used.

Given the magnitude of Winstein's contributions to chemistry and the recognition he had already received for them, this lack of generosity to a young and unknown chemist was uncharacteristic and unnecessary. It also had unfortunate consequences for organic chemistry because the large majority of "nonclassical carbocations" are, in fact, π complexes, and their chemistry can be interpreted much more simply and effectively on this basis[15] than it can in terms of the obscure "dotted line" representation (**26**) that Winstein introduced and which, thanks to his influence, still remains in general use.

The strengths of dative bonds are limited by the charge transfer involved in them. Consequently, simple olefin π complexes are stable only if the apical group is a very strong electron acceptor. However, if the apical group has p or π electrons, these can be used to form a second dative bond, opposite in direction to the first, in which the empty antibonding π MO of the olefin acts as the acceptor. Figure 1 indicates the notation used to depict this bond. Such back-coordination greatly strengthens the π complex because the charge displacements in the two dative bonds are in opposite directions. Strong bonding can thus occur without any large net transfer of charge. Indeed, the MO description of such a situation is equivalent to that in a three-membered ring. Thus, in principle, there should be a continuous transition between pure π complexes, without back-coordination, and microcycles. (A further discussion is given in references 15 and 16).

26

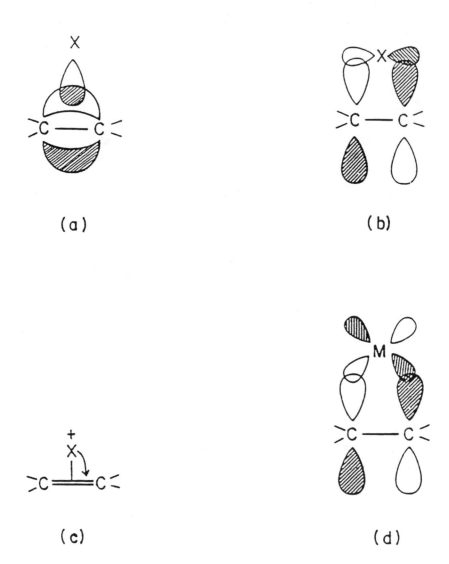

Figure 1. (a) Overlap of orbitals in a π complex; (b) back-coordination in a π complex where the apical group has p or π electrons; (c) symbol for π complex with back-coordination; (d) back-coordination using a filled d AO of a transition metal in an olefin π complex.

In the discussion at the Montpellier conference, I was asked if metal–olefin complexes could be interpreted as π complexes. The structures of such species, in particular of the anionic component $(Cl_3PtC_2H_4)^-$ of Zeise's salt, were, at that time, a subject of much interest because they clearly involved a kind of bonding with no precedent elsewhere. I pointed out[17] not only that these species could, of course, be formulated as π complexes, but also that ones derived from transition metals should be stabilized by back-coordination, a filled d AO of the metal acting as the donor (Figure 1d[17]).

Chatt had indeed found evidence for back-coordination in such complexes[18] but had explained it in a different way. He assumed that the ethylene in them is not present as such, but rearranges to ethylidene, $CH_3CH=$, the complexes being complexes of carbenes rather than olefins. Thus, he formulated the ion in Zeise's salt as $(Cl_3Pt=CHCH_3)^-$. He and Duncanson set out to distinguish between the two structures by using the then-novel technique of IR spectroscopy. The IR spectrum of Zeise's salt conclusively supported my π-complex formulation. Chatt subsequently carried out extensive studies of analogous complexes that he naturally interpreted as π complexes. As a result, many still attribute the π-complex structure of metal olefin complexes to him, and the theory is still commonly referred to as the Dewar–Chatt–Duncanson theory.[18] π-Complex theory now plays a major role in organometallic chemistry, making it all the stranger that organic chemists still ignore it.

How did it come about that I was asked to give a lecture at a major international conference even though, apart from a year at Oxford as an ICI fellow, I had not at this point officially carried out any independent research? The reason I was asked is that I was already very well known as a result of my moonlighting activities. As I have already mentioned, the stipitatic acid problem had baffled all the leading organic chemists in Britain and America (including Woodward and Robinson!), and my solution aroused even more interest because, as I pointed out, it applied also to colchicine, another and much more important compound for which no satisfactory structure had been suggested. My suggestion, that both contained an entirely novel aromatic system, which I termed tropolone, started the study of nonbenzenoid aromatic chemistry. Similar comments applied to my π complex theory, which was also entirely novel, and to my first book, which began the conversion of organic chemists to MO theory. It would be difficult, if not impossible, for a young chemist today to make a similar entry into the world of organic chemistry because it is most unlikely that any totally novel structures or phenomena remain to be discovered. Furthermore, the establishment today, in Europe as well as America, is geared to preventing the chemical boat being rocked by young mavericks.

The Move to Maidenhead (1945–1951)

When the war ended, Imperial Chemical Industries (ICI) established a number of postdoctoral fellowships in chemistry at universities in Britain, and I was awarded one at Oxford. Able for the first time to carry out my own research, my first step was to try to synthesize tropolone. I decided to try all the methods I could think of, 14 in number, to see if any would work the first time. I was planning, if they all failed, to return to the more promising ones and give them intensive treatment. All, naturally, did fail. However, I had, in the meantime, become interested in the benzidine rearrangement and decided to carry out kinetic measurements to test my suggested mechanism, intending to return to tropolone later. Although the work on the benzidine rearrangement was published,[19] I never did return to tropolone because at this point a completely unexpected series of events diverted me from the normal academic career I had planned.

It began with an invitation from the Oxford University Press to write a monograph on theoretical organic chemistry. This was clearly an exciting and challenging opportunity. However, I soon realized that I did not know nearly enough about

51

physical chemistry or quantum theory to do the job properly. Just at this point, I was offered a position as a physical chemist to carry out basic research in a new laboratory that Courtaulds, Ltd. (manufacturers of rayon and acetate fibers) was setting up in Maidenhead, a small town near London.

What an opportunity! Not only a chance to learn physical chemistry but actually to be paid to do it! I naturally accepted, and that was the end of tropolone. As a postscript, every single approach to tropolone that I tried has since been made to work, and, until quite recently, no route to tropolones had been reported that I had not tried. The lesson is obvious and one that I have taken to heart ever since. *Nothing ever works the first time in organic chemistry, but almost everything does eventually if you try hard enough.*

My move to Maidenhead was very unorthodox and shocked Robinson, but then my whole career has been unorthodox. Because of my early lone start in chemistry, and because of the way teaching was conducted at Oxford, I was almost entirely self-taught. Because of the war, I was deprived of a normal entry to academic life, and my move to Courtaulds further isolated me from the academic community. When I finally joined it, I did so at the top, bypassing the usual academic ladder by moving from industry to a position at least equivalent to a named chair at a leading American university. This unusual sequence of events is probably why I have never had any respect for accepted ideas simply because they were accepted, or for people simply because of the positions they held. Two things I did inherit, from Robinson, were a passion for absolute integrity in science and a loathing of "operators," of whom it sometimes seems chemistry has more than its share.

I spent 6 years at Maidenhead, working with Bamford on a mixed bag of problems. Bamford, or Bam as he was called by friends, was head of the Physical Chemistry Division at the Maidenhead laboratory. I should perhaps explain that in England at that time, among men, first names were used only by relatives and close personal friends. Indeed, in many cases, such as Bam's, first names were not used at all, being replaced by nicknames. The normal semi-intimate form of address was to use the last name, without a qualifying title. To use first names incorrectly was considered most improper and grossly impertinent. No wonder visiting Americans often found the atmosphere chilly when they blithely first-named everyone!

Bam and I were the first to measure absolute rate constants in a vinyl polymerization,[20] and we were also the first to measure absolute rate constants for an autoxidation.[21] The latter followed from studies of an amusing technical problem: the tendering of fabrics by light. Fabrics dyed with certain vat dyes, in particular yellow ones, fall to pieces on exposure to sunlight, a phenomenon that was causing problems for manufacturers of curtains. We concluded[22] that the reaction involved either extraction of hydrogen atoms from the fiber by photoexcited dye molecules, which led to radicals that reacted with oxygen, or, in the presence of base, to oxidation of hydroxide ions to radicals by electron transfer, the photoexcited dye acting as an electron acceptor. These suggestions were novel at the time, and we used one of the dyes as a photoinitiator in our kinetic studies[21] of the autoxidation of tetralin.

A third line of work was concerned with the mechanism of the thermal decomposition of acetic acid to ketene and water, a reaction that formed the basis of a then-new process that Courtaulds had adopted for making the acetic anhydride they needed to manufacture cellulose acetate. We found that the decomposition of acetic acid can take place in the gas phase by either of two homogeneous first-order reactions, one leading to ketene and water and the other to methane and carbon dioxide.[23] Very little seems to have been done since on either of these reactions, curious in view of their obvious interest in connection with theories of pericyclic processes.

Monograph on Theoretical Organic Chemistry

During the first part of this period, I was also working on my book at night and on weekends. When Robinson first heard about the book, he suggested that we collaborate on the project, a suggestion that, of course, I welcomed. I discovered later that this was the 19th time this had happened with someone in his department. Each of the first 18 authors waited for Robinson to do something about it, which he never did, so none of the books ever appeared. Knowing Robinson, it never occurred to me that he would be able to find time to write anything himself, so I went ahead on my own, and sent him each chapter as soon as it was finished for revision and comments. However, when the book came to its end, Robinson still had not been able to read any of it, so after 6 months, with some encouragement from The Oxford University Press, he decided to withdraw from the project, apart from contributing a very nice preface.

When *The Electronic Theory of Organic Chemistry*[12] appeared in 1949, it was something of a landmark. It was the first general account of organic chemistry in terms of MO theory. MO theory is essentially a method of computation, like valence bond

THE

ELECTRONIC THEORY

OF

ORGANIC CHEMISTRY

BY

M. J. S. DEWAR

Two pages from Leopold Ruzicka's copy of my first book. When I saw these pages recently, I felt surprised and happy to see how much interest he had in it.

VIII   REACTIONS OF CARBON—CARBON MULTIPLE BONDS   143

does not explain the exclusive *trans* addition of halogen to olefines. The intermediate cation would be symmetrical and should give a mixture of *cis* and *trans* adducts.

Roberts and Kimball[†] pointed out that the latter effect could be explained if the intermediate ion had a cyclic structure, the subsequent reaction with the anion being an S_2 replacement in which the ring opens and the *trans* adduct is formed:

The exclusive anionoid reactivity of olefines can then be explained if the intermediate 'cyclic' cation is in fact a π-complex in which a bromous cation is linked to the π-electrons of the $C=C$ bond (cf. p. 17). Its formation will be electronically analogous to the formation of a quaternary ammonium salt from a tertiary amine and alkyl bromide:

cf. $R_3N + R-Br \longrightarrow R_3\overset{+}{N}-R\ Br^-$.

The corresponding reaction with an anion is impossible since ethylene has no vacant electron orbital of low energy and cannot therefore act as an electron acceptor without actual fission of the π-bond. (The latter process does occur in reactions with atoms or radicals; Chapters XIII, XIV.)

Further evidence is provided by the orientation of the products from the addition of iodine chloride to unsymmetrical olefines. These reactions resemble the normal halogen additions and the intermediate cation is presumably formed by addition of the less electronegative halogen to the olefine:

$$\overset{CH_2}{\underset{CH_2}{\|}} \quad I-Cl \longrightarrow (C_2H_4I)^+\ Cl^-.$$

Now if the cation were a simple cyclic compound, as Roberts and Kimball suggest, the subsequent S_2 replacement should take place at the end of the olefine most positively substituted, since S_2 replacements are accelerated by positive substituents, and the chlorine should appear there in the products. Actually the observed orientation is just the opposite. This result can be explained in terms of the π-complex theory. In the π-complex substituents will polarize the double bond; therefore in an unsymmetrical olefine the I+ cation will be attached unsymmetrically, being closer to the

[†] *J.A.C.S.* 1937, 59, 947.

(VB) theory. Neither provides the kind of qualitative picture that chemists need in thinking about molecules, reactions, syntheses, etc. Pauling had provided such a picture on the basis of VB theory; I had now done the same in terms of MO theory. Although older chemists, reared on resonance theory, were not impressed, and although some wonderfully scathing reviews appeared as a result, the book sold well and clearly did make a major impression on younger chemists. It undoubtedly started the conversion of organic chemists to MO theory.

The problem with resonance theory, apart from doubts[24] concerning the validity of the form used by organic chemists, is that it is too qualitative and intuitive. One can explain facts once they are known but rarely make reliable predictions. The same was true of the treatment in my book. The key to solving this problem was provided by Coulson and Longuet-Higgins[25] while my book was gestating, through their realization of the potential of perturbation theory in this kind of connection and also of the significance of the Coulson–Rushbrook pairing theorem,[26] a series of simple relationships between the MOs of alternant hydrocarbons (AHs). In this way, they were able to deduce a number of simple relationships between different conjugated systems, exactly the kind of thing needed as the basis of a general treatment of chemistry.

However, one link was still missing. In their treatment, the perturbations involved were second-order ones, involving a knowledge of the coefficients of AOs in all the MOs involved. These can be found only by using a computer. A treatment of this kind would clearly be useless as the basis of a general chemical theory to serve as a replacement for resonance theory.

Reilly Lectures

In 1951 I was asked to give the Reilly lectures at Notre Dame University, a very exciting invitation. Unfortunately, I had to go alone. Mary had to stay home because we had nobody to look after the children and we could not have afforded to take them with us. This series at the time involved a full 6-week course of lectures. I decided on MO theory as my topic, intending to base my lectures largely on the new work by Coulson and Longuet-Higgins.

While preparing the lectures, I suddenly saw the vital missing link. Coulson and Longuet-Higgins had considered only the changes when two *even* AHs, that is, ones with even numbers of conjugated carbon atoms, unite to form a larger one. What about the union of *odd* AHs? According to the pairing theorem, the MOs of an AH occur in pairs, with equal and opposite energies relative to that of a carbon $2p$ AO. The number of MOs in an odd AH is, however, odd, so one MO must remain unpaired. It therefore has the same energy as a carbon $2p$ AO. All such nonbonding MOs (NBMOs) of odd AHs consequently have similar energies in the HMO approximation, a result that indeed forms part of the pairing theorem.[26]

When two odd AHs unite, there is a large interaction (first-order perturbation) between their NBMOs because their energies are identical. Furthermore, if the odd AHs are neutral, they contain an odd number of electrons and are consequently radicals (e.g., allyl or benzyl) in which the NBMOs are singly occupied. In the product of union, *both* the nonbonding electrons can go into the lower of the MOs derived from the interaction between the NBMOs, leading to a first-order perturbation. Because first-order perturbations are generally much larger than second-order ones, the latter can be neglected in comparison with first-order ones. The first-order perturbation is, moreover, not only simpler, involving only a single term, but it also involves the coefficients in just *one* MO, the NBMO. As Longuet-Higgins had already shown[27] on the basis of the pairing theorem, the coefficients of AOs in an NBMO can be found in no time at all by a simple pencil-and-paper calculation.

This was the clue! By using it, I was able to develop a complete theory of organic chemistry, completely free from subjective elements and even providing semiquantitative estimates of many of the quantities (e.g., heats of reaction or activation) needed to interpret chemical reactivity. Because I had developed this for my Reilly lectures, it seemed appropriate to publish it from Notre Dame, so I submitted it to the *Journal of the American Chemical Society* (*JACS*) in the form of a series of six papers.[28] To emphasize the rigorous basis of what I later termed[29] perturbational molecular orbital (PMO) theory, I presented it in the form of 74 theorems with formal proofs of each. Because I had been warned that *JACS* was fierce about space, I also took care to write the papers very compactly.

First visit to the United States. Lecture at Jackson Laboratory, Dupont, May 7, 1951, "Recent Developments in Theoretical Organic Chemistry." Left to right, D. G. Ebenhack, Dewar, M. Hunt, and H. E. Schroeder.

The papers went to three referees. All of them agreed that the author had to be mentally defective to produce such stuff, but that the papers should certainly be published provided they were shortened by 25%. Being young, I meekly complied, so the papers in their final form were very difficult to read. I only later discovered that the *Journal of the American Chemical Society* had at that time a firm policy of publishing *no* purely theoretical papers. Mine not only appeared but led to the ban being rescinded.

University of London, Queen Mary College (1951–1959)

I had not expected to stay so long at Courtaulds, but this was a period when there were no academic openings in Britain. When the break came, it did so in a rather unexpected way. I was offered not only an academic position, but a chair at the University of London, at Queen Mary College (QMC). This was a surprising appointment, given that I had no previous academic experience, given that the University of London ranked next to Oxford and Cambridge in the pecking order, and given that I was the youngest person (age 33) ever to be appointed to a chair in London (apart from one or two mathematicians).

The move to London totally changed our lives. Our married life until then had been rough, particularly for Mary. In normal times, she would undoubtedly have received a fellowship at one of the women's colleges in Oxford as the flying start to an academic career. The war stopped this. All women in Britain were drafted. However, Mary was appointed to a responsible position in the Ministry of Town and Country Planning, which could have been the start of a good career in the Civil

Chemistry Department at Queen Mary College, 1954, middle center. After I left, this Victorian monstrosity was torn down—with difficulty; the walls were more than three feet thick—and replaced with a beautiful modern building with 11 floors and wonderful views over London. This relic had five floors and no elevator. My office was on the fifth floor.

Service or a stepping stone back to academic life after the war. In Oxford, however, no suitable jobs were available, so she ended up in a minor job in a local government office until she was rescued by having our first child.

Maidenhead was even worse. We lived in a perfectly horrible house that was moreover almost 2 miles from the nearest shops. Because there were still no buses, because maids and cars had vanished with the war (gasoline was available only for essential purposes), and because I was working crazy hours for the first 3 years, writing my book as a sideline from a very exacting job, Mary had a very rough time. She not only had to look after the house but also do all the shopping, and because babysitting in Britain was still in the distant future, she had to shop pushing two very obstreperous babies in a baby carriage. Furthermore, because refrigerators were not available (all appliances, even electric irons, had vanished with the war), she had to shop every day, and shopping at that time meant hours spent in queues, the few things not rationed being in extremely short supply. Later, when we were able to get a maid and a car and a refrigerator, Mary got a part-time lecturing position at the Maidenhead Technical Institute, but finding time for this took heroic efforts on her part.

To make things worse, in 1948 I was suddenly crippled. It struck like a bolt from a blue sky. I had been playing in a bridge tournament in Birmingham and set off to collect my car from a nearby garage. When I got back to the hotel to collect my luggage, I could hardly move. I was in agony and I remained that way. When I got home to our doctor, she had my spine X-rayed, suspecting a damaged disc, but the X-rays showed nothing. At this point, I had one of my few pieces of sheer good luck. Our doctor's brother, Wiles, was a leading orthopedic surgeon at the Middlesex Hospital, one of the major hospitals in London, and one of the leading authorities in the world on spinal conditions. He had told our doctor about a rare congenital condition (spondylolysisthesis) that had just been recognized and that frequently failed to show up in regular X-rays. So she sent me to see him—and sure enough that was my problem. The little bones holding a pair of spinal vertebrae together are missing. As long as the vertebrae stick together by friction, all is well, but once they begin to move, they pinch the spinal column and produce agonizing pain. The only solution at that point is to fuse the vertebrae together with a bone graft, a very tricky operation that Wiles had pioneered. In my case, one vertebra was attached neither to the one above it nor to the one below it, so I needed a double spinal fusion. Wiles duly operated on me. Fortunately, neither Mary nor I knew at the time that the success rate for the operation was still very low. Indeed, Wiles was one of the very few who had had any success at all. I was one of the lucky ones in the sense that I have been able to function more or less normally since. However, walking or standing for any length of time since then has always caused me major pain.

This condition led to a big change in my life. I had previously been extremely active, taking part in games of all types, running, ice skating, and rock climbing, with great enthusiasm. Since 1949 I have been barred from everything but swimming. I must admit that my skill never matched my enthusiasm. I was

Leading a rock climb in North Wales in 1942, Holly Tree Climb, Idwall. This photo occupies an honored place on my desk.

in fact hopelessly incompetent at almost everything. I still remember an occasion when it was thought, to everyone's amazement, that I had won a 1-mile race, until it was realized that I was in fact a lap behind the winner. However, I enjoyed trying, and it was hard to be deprived.

In London, we found a beautiful house on the edge of Richmond Park. By this time, life was beginning to return to normal in Britain, so Mary decided to take a belated Ph.D. at the University of London. As she knew none of the historians in London, she naturally decided to register for the Ph.D. at Queen Mary College to avoid paying fees. So she studied with Bindoff, the professor of history there.

Nowadays, this would be considered completely normal. In 1951, it caused a sensation. Mary was indeed a pioneer. It was unheard of then for wives to return to work, let alone to return to student life at a university to take a degree. They were expected to stay at home and look after their families. Furthermore, the University of London suffered even more than most of the rest of Britain from social snobbery and the British class system. Everyone in London was expected to show proper deference to those higher in the pecking order and, conversely, conscious superiority to those below. Professors were of course at the top and students at the bottom. Imagine then the predicament of a professor supervising a graduate student who was not only the wife of another professor but also the wife of a colleague! Bindoff never really succeeded in coping with the situation. Equally, the other students at first regarded Mary with suspicion and were wary of her. It took her a lot of effort to thaw them out.

The hierarchical system in British universities was a sore trial to us. It was no consolation to us that professors were in the upper reaches. We felt the whole thing ridiculous. The situation was made worse by the fact that we were much younger than our official equals. Those our age were the equivalent of assistant professors in the United States. I remember one occasion when Mary and I had got on first-name terms with one of the younger members of my chemistry staff (as the faculty are described at universities in Britain). The principal of Queen Mary College cornered him one day and said, "I know that Professor and Mrs. Dewar have asked you to address them by their first names. However, if you are thinking of your

The Doctors Dewar in London, 1956, the day Mary got her Ph.D., outside our house in London, in East Sheen, on the edge of Richmond Park.

career, you would be well advised not to do so in public"! I might add that the person in question is now a fellow of the Royal Society and one of the most eminent chemists in Britain.

One unfortunate consequence of the way chance operated for Mary was that she got entrenched in English Tudor history because that was Bindoff's field. This was bad luck because 16th-century English history is a bit esoteric in America, and Mary had had no previous special interest in it. Almost any alternative would have been much better for her. Whereas the books she has written have been very well received by other

Tudor historians, there are not many Tudor historians in the United States.

Life also improved for us in other ways with the move. We were able to go to London theatres and did so avidly. Rationing at last ended. We now had a rapidly growing circle of friends. As was usual in our circles in Britain at the time, our sons Robert and Steuart were soon of the age to go to a boarding school and went to my old preparatory school, Copthorne. Robert had the unexpected experience on his first day there of being taken to see a desk with the memorable phrase "Dewar is a fool", which had been carved on it 30 years earlier.

In those days, there was only one professor in each department in universities in Britain, and his power was absolute. So here I was, expected to run a department, not only with no experience but also with no help, because the two next senior members of the chemistry staff, Jones and Hickinbottom, who were both near 60 and should have been my right-hand men, had not spoken to each other for 4 years and agreed only in resenting my arrival, each of them thinking that he, himself, should have had the job. I was, therefore, unable to delegate anything to anybody without trouble. If I delegated anything to Jones, Hickinbottom was up in arms, and vice versa. If I delegated anything to anyone else, both of them were up in arms. I did succeed in intimidating them sufficiently for them to attend faculty meetings without misbehaving, but that was as far as I ever got.

Furthermore, the department had no equipment and no money. It had been stripped during the war, and my predecessor, Partington, who did not even live in London but traveled in from time to time from Cambridge, had made no attempt to re-equip it. The annual budget for chemicals and equipment, for teaching as well as research, was the equivalent of $7000. At that time, there were no outside sources of research support. It was rough.

The first step was to get equipment and more money. Nobody else could have done this. I did it by an approach so contrary to convention that nobody at Queen Mary College knew how to meet it. I simply overspent my budget. I was of course helped by the fact that professors in Britain had dictatorial powers and by the tacit assumption that none of them would ever think of breaking rules. However, I could not have done it without the aid of my laboratory steward, who was a financial genius *manqué*. He should really have been on the stock exchange. If he had been, I am sure he would have become a multimillionaire. To spare him possible embarrassment if he is still alive, I will refer to him as X.

My first move at QMC was to order a huge amount of basic laboratory equipment. X was horrified. "But, Sir," he said, "we haven't got the money." I replied, "True, but we have to have the stuff." After the first shock had subsided, X took to the project with gusto, and at the end of the year it turned out that we had overspent our budget by 100%. The college was horrified. Nothing like this had ever happened before in its entire history. The principal explained to me the horror of what I had done, but he was very sympathetic and arranged for the chemistry budget to be doubled the following year. At this point, I summoned X and said to him, "In future we will overspend our budget each year by exactly 100%. More than 100% would be unreasonable, and we do not want to be unreasonable, but less would be ridiculous."

The QMC administration was naturally perturbed by the situation, and its accountants scrutinized our accounts each year in the utmost detail to prevent us from overspending again. However, they were no match for X. Each year, when the smoke had cleared, it turned out that we had overspent our budget by between 99% and 100%, and each year my budget for the following year was doubled. The climax came after 5 years. At this point, our budget had at last become adequate, being 32 times what it had been originally. However, my other major problem, the vendetta between Jones and Hickinbottom, was beginning to get me down, and my efforts to get QMC to appoint a second professor, on whom I could have dumped the problem, had failed for lack of money.

Soon after the fifth budget debacle, I happened to be sitting next to the principal at lunch one day when he turned to me and said, "Wouldn't it be a help to you if we appointed a second professor in chemistry? It would make life so much easier for you if you could hand over the budget, and all the other routine problems that cause you such trouble, to someone else." I replied, "What a wonderful idea!" So we appointed Keble

Sykes as the second professor, and at the end of the year our budget was exactly balanced. The principal always thought he had brought off a masterstroke. In fact, it was I who put the idea subliminally in his mind at that lunch. I will forever be grateful to X. I owe him much.

Keble's appointment came at just the right time for me because at this point I was invited to spend a semester at Yale as visiting professor. Now I could leave Keble in charge as acting head of the department. At the first staff meeting he held after I left, Jones and Hickinbottom came to blows and had to be separated.

I should perhaps add that I am hopeless at administration. I am good at seeing what has to be done but hopeless at getting it done. A good administrator operates by manipulating people. I have never been any good at this; my effort with the principal at QMC was just a rare exception. In my 8 years at QMC, I dragged the chemistry department from the bottom of the heap at London University to the top, but I did so only by brute force and at the expense of much unnecessary bloodshed. However, my failings as an administrator, and my recognition of those failings, have made me a very good judge of administrative ability in others.

So while I was now free for the first time, at the age of 33, to work on my own projects, anything I did had to be in the context of very limited facilities. Nevertheless, we got a lot done. At that time, there was no movement at the graduate level. Students took Ph.D. degrees at the same university as their B.A. or B.Sc. I was fortunate in that London got the cream of students in Britain, next to Oxford and Cambridge, so right from the start I had very good ones in my research group. Still better, one of the first members of it was Roly Pettit, who came from Australia with an 1851 fellowship, officially to work for a Ph.D., although he had already a Ph.D. in Australia. After taking a second Ph.D. in London, he got an ICI fellowship so, in all, he was with me for 5 years.

My group at QMC was surprisingly productive, considering the lack of facilities and the amount of my time taken up by my own responsibilities as head of the department. Admittedly, it took some time for us to get started, given that there were not even stools in the laboratories, let alone things like glassware

The first Dewar research group at Queen Mary College, 1953. Seated left to right, R. Pettit, Dewar, and L. C. F. Blackman. Standing, left to right, J. M. W. Scott, P. M. G. Bavin, J. S. Dave, D. Morley, and E. T. H. Warford.

and vacuum pumps. The papers I published in the first 2 years were therefore mostly theoretical.

The most interesting publication was a joint one with Christopher Longuet-Higgins[24] in which we explained why resonance theory works. As the Pullmans had shown some years before, the basic assumption of resonance theory, that is, that the dominant contribution comes from unexcited (classical) structures, has no basis in VB theory. Christopher and I showed that there is an accidental correspondence between the numbers of resonance structures that can be written for an odd AH radical (e.g., allyl or benzyl) with the unpaired electron located at some atom x, and the coefficient (a_{ox}) of the corresponding AO in the NBMO of the radical (Chart I). This correspondence leads to a parallel between the predictions of PMO theory and resonance theory that is responsible for the success of the latter. The correspondence fails, however, in compounds containing $4n$-

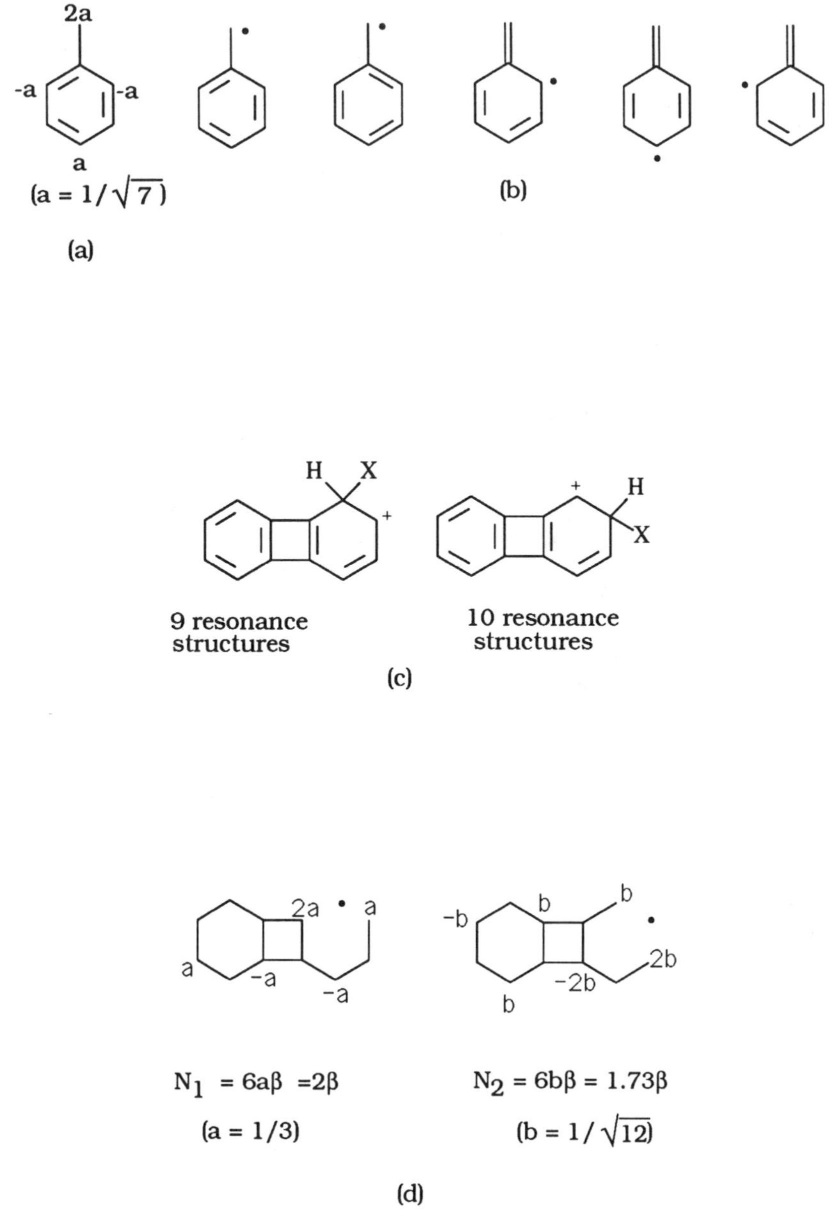

(a)

(b)

9 resonance
structures

10 resonance
structures

(c)

$N_1 = 6a\beta = 2\beta$

$(a = 1/3)$

$N_2 = 6b\beta = 1.73\beta$

$(b = 1/\sqrt{12})$

(d)

Chart I. (a) NBMO coefficients in benzyl; (b) resonance structures for benzyl radical; (c) resonance structures for Wheland intermediates in substitution of biphenylene; (d) NBMO coefficients in these intermediates.

membered conjugated rings, and resonance theory consequently gives wrong predictions in such cases. For example, resonance theory predicts biphenylene (27) to undergo electrophilic substitution in the 1-position (Chart Ic), whereas PMO theory correctly predicts the 2-position to be the more reactive (Chart Id).

Christopher is one of my oldest friends. We were both scholars at Winchester, though 4 years apart, and we both went to Oxford, to Balliol College. One of my photographs reminds me of the only time there was friction between us, when another friend and I took Christopher to North Wales, to introduce him to rock climbing. One day, I was leading quite a hard climb and on the hardest bit, near the top of a long pitch, I was nearly pulled off by a great jerk on the rope. It turned out that Christopher had succeeded in getting the rope between me and him, and between him and the third man, inextricably entangled. I was left holding on virtually by my fingernails for what seemed like hours while they untied themselves and disentangled the rope, inch by inch, so that I could go on. I must admit that there was a momentary stress in our friendship when we all got to the top of the pitch! I think Christopher was wise to become a theoretician.

Christopher's defection from chemistry to study the mode of action of the brain was a great loss to chemistry, but understandable because he has never had proper recognition for his contributions. Probably the worst example concerned boron hydride chemistry. While he was still an undergraduate at Oxford, Christopher solved a problem of major topical interest by establishing the structure of diborane, an electron-deficient molecule that cannot be represented in terms of normal covalent bonds because there are too few electrons. In one of his weekly essays for Ronnie Bell, Christopher showed, by analysis of its infrared spectrum, that diborane has the now accepted bridged structure. This had been suggested as a possibility, but the

27

suggestion had been rejected because Bauer, one of the leaders in the electron diffraction area, had concluded from an electron diffraction study that it had an ethanelike structure. This conclusion had moreover seemed established because Pauling had suggested a possible interpretation for the ethanelike structure in terms of resonance theory, whereas no interpretation of the bridged structure had been suggested.

A few years later, Christopher solved this problem also, in terms of a novel type of bond whose existence he deduced on the basis of MO theory and which he used to interpret several other electron-deficient molecules. The structures of most boron hydrides and related compounds, e.g., the carboranes, can be interpreted in terms of such three-center two-electron bonds. Christopher also discovered the only other novel type of bond encountered in the boron hydrides and carboranes, from a detailed MO study of the dodecaboride anion. Thus he alone is responsible for the theory of bonding in the boron hydrides and related carboranes, as well as having been the first to establish the structure of one of them. Subsequent workers have merely applied Christopher's ideas and, in some cases, acquired misplaced credit for the ideas themselves.

Climbing holiday in North Wales, the one where Christopher (on my right) almost terminated my chemical career. John Mills is on my far left.

Localization of Bonds

Another early study with Roly Pettit was concerned with the localization of bonds. One of the problems in MO theory is its inability to explain why bonds in molecules frequently seem to be localized. The MOs of a molecule include contributions by AOs of all the atoms in it (apart from limited restrictions due to symmetry). Although numerous attempts have been made over the years to explain bond localization by transforming the MOs of a molecule into orbitals mainly composed of AOs of pairs of atoms, and although this transformation can often be done quite effectively, the resulting localized orbitals are no longer independent.

The energy and other properties of a molecule cannot, therefore, be written as sums of corresponding localized bond energies. There are contributions from the interactions between the localized orbitals. I had already shown why classical structures play a role in conjugated molecules, in the original PMO papers.[28] Roly and I extended this kind of analysis to paraffins.

Professor H. C. Longuet-Higgins, one of my oldest friends. Christopher has never had the recognition that he should have had for his many original and extremely important chemical contributions. It was a great loss to chemistry when he transferred his talents elsewhere.

H. C. Brown of Purdue University won the Nobel Prize for his work on boron hydride chemistry. He was one of the main protagonists in the famous, or infamous, controversy concerning the structure of the 2-norbornyl cation. After many years, in response to new evidence, I changed sides, the first and only participant to do so, and much to Herb's satisfaction!

We were able to show[30] that if one starts with a molecular wave function corresponding to localized two-center bonds, the interactions between them are additive and can therefore be absorbed into empirical "bond energies", sums of which will reproduce the heats of atomization of molecules.

The localized bond model is therefore no more than a model. The bonds in molecules are not "really" localized, but that doesn't make the model any less useful. This situation indeed illustrates a basic principle I learned from reading Collingwood, that is, that the criterion of a scientific theory is not its truth but its success. Whether a theory is "true" is totally irrelevant. We cannot tell, on the basis of pure reason, what the universe is or why it is. The most we can do is find out how it works. Because we do not, and cannot, understand nature, we try to predict its behavior from that of models. A model is a simple mechanism that reproduces the behavior of a more complex mechanism well enough for the behavior of the latter to be predicted from that of the model. Any scientific theory is the specification of a model. If a model fails, we either improve it or replace it with a better one.

From this point of view, fanatical belief in the "truth" of any theory becomes ridiculous. One does not believe in the "truth" of a model. Deeply held beliefs of any kind are incompatible with science. They serve no useful purpose and they obstruct progress. A good example is the classic controversy concerning the 2-norbornyl cation. The almost religious fervor with which this has been conducted reminds one of the early schisms in the Catholic church. I was, and remain, the only participant in this rather unfortunate affair to change sides, in response to new and, to me, convincing evidence.

R. S. Mulliken at Chicago University won the Nobel Prize for his contributions to quantum chemistry. Officially a physicist, Robert was in the Physics Department. As a result, although the chemistry and physics buildings were next door to one another, we never met in Chicago except at cocktail parties. Being in different buildings is a bigger obstacle to visiting than being in different countries!

These ideas concerning the nature of bond localization led some years later to a reevaluation of conjugation and hyperconjugation. At that time, Pauling's interpretation of species, such as butadiene, in terms of resonance interactions between adjacent multiple bonds, and Mulliken's views concerning hyperconjugation, were generally accepted. It occurred to me that many of the effects attributed to conjugation or hyperconjugation in such systems might in fact be due simply to the different hybridization of the AOs involved in forming the different kinds of C–C single bonds. Thus the central bond in butadiene is formed by sp^2 carbon AOs whereas C–C bonds in paraffins are formed by sp^3 AOs. Indeed, all the observed differences could be explained quantitatively in this way.[31,32]

Although the picture we presented was oversimplified, it led to badly needed rethinking of basic ideas in the area. As an amusing postscript, I was not only invited to an International Symposium on Hyperconjugation at the University of Indiana, but also gave the first lecture at it. The Symposium thus started with a lecture claiming that the subject of the symposium did not exist! An expanded version of my lecture appeared soon after in the form of a small book.[33]

Some Early Projects

PMO theory represented a major advance over resonance theory in that it led to definite and quantitative predictions.[28,48] My first major experimental project at QMC was to test it. I chose as the first test case electrophilic substitution in polycyclic benzenoid hydrocarbons, because here the predictions of PMO theory are unambiguous, compounds of this kind being even AHs. Those unacquainted with PMO theory will find a brief outline in Appendix B. Curiously enough, no one had systematically measured the rates of such reactions. Indeed, no one has repeated our work since, apart from measurements for one or two specific compounds. This is very surprising, given the obvious potential theoretical interest of such a study and the fact that it could now be carried out very easily by using GLC. We had to use much more cumbersome procedures, GLC having not yet been invented. Our results[34,35] were in very good agree-

Conference on hyperconjugation at Indiana University, June 2–4, 1958. I gave the first lecture, arguing that hyperconjugation did not exist. Front row, left to right: P. B. D. de la Mare of University College, London University; E. Berliner of Bryn Mawr College; D. A. McCaulay of Standard Oil Company; A. Streitwieser of the University of California—Berkeley; R. T. Arnold of the University of Minnesota; and L. E. Sutton of Oxford University. Back row, left to right: R. S. Mulliken of the University of Chicago; F. T. Gucker of Indiana University; E. Campaigne of Indiana University; E. S. Lewis of Rice University; Dewar (of Queen Mary College, London University); H. Shull of the University of Colorado; W. M. Schubert of the University of Washington; S. Winstein of the University of California—Los Angeles; and J. Shiner of Indiana University.

ment with the PMO predictions, and later we obtained similar agreement from studies[36] of substitution in heteroaromatic compounds and of the solvolysis of arylmethyl chlorides.[37]

Another early project was suggested by my work on tropolone. Since publication of my original papers, it had become clear that the aromaticity of tropolone is in fact due to its being a derivative of the Hückel $C_7H_7^+$ polymethine cation, **28**, which had been termed tropylium. Roly Pettit set out to make it. We avoided the obvious route from cycloheptatriene by halogenation and dehalogenation because I knew Bill Doering at Yale

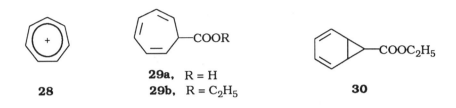

28

29a, R = H
29b, R = C$_2$H$_5$

30

was trying that. Instead, we decided to try oxidative decarboxy-lation of cyclohepta-1,3,5-triene-7-carboxylic acid (**29a**). Büchner had reacted benzene with ethyl diazoacetate and obtained a product that he formulated as **30**. The corresponding acid, on treatment with alkali, isomerized to a species (X) that he formu-lated as **30**.

Roly duly prepared X according to Büchner's recipe and tried to convert it to tropylium in a number of ways, all of which failed. However, we omitted one obvious method, namely Curtius degradation, because Alan Johnson, at Notting-ham University, was trying this. Alan's coworker was, to say the least, not very competent. It took him a year to obtain a

Tetsuo Nozoe visiting Queen Mary College, about 1956, with me and Roly Pet-tit. Nozoe used to visit us whenever he was in London because of our common interest in tropolones.

urethane, $C_7H_7NHCOOC_2H_5$, which hydrolyzed very easily to give urethane itself and some other unspecified product. Alan assumed the parent urethane to be **31**, attributing its hydrolysis to easy heterolytic cleavage to tropylium. To us, however, the ease of hydrolysis suggested that it was in fact **32**, an enamine

derivative, and that the parent ester (**30**) had been wrongly formulated.

Alan having told me that he planned no further work on the problem, I let Roly loose on it. In 48 hours he had repeated the preparation and hydrolysis of the cycloheptatrienylurethane, and obtained not only urethane but also cycloheptadienone (**33**). So we were right! Alan's compound *was* **32** and Büchner *had* misassigned the structures of his acids. It now seemed likely that his original adduct was in fact **29b**, not **30**, as is indeed now known to be the case.

So Roly had been making the compound we wanted and then carefully isomerizing it to something else! When he repeated our attempted syntheses, using the right acid, every one of them worked.[38] However, the year's delay let Bill Doering scoop us, by the rather unfair expedient of reading the literature and finding that **28** had in fact already been made, half a century earlier, by Meerwein, who obtained the chloride without realizing what it was.

Another major project started by chance. A student, who had taken a M.Sc. in India, came to QMC to work for a Ph.D. in heterocyclic chemistry. When he came to talk to me, he said, "My work for my master's degree involved liquid compounds. I would like, if possible, to continue working on liquids since I have no experience of crystalline solids." Being in an inventive mood that day, I immediately suggested that he try to synthesize boron-containing aromatic compounds, the idea being that because B^- and N^+ are isoelectronic with C, one should be able to replace two adjacent carbon atoms in an aromatic ring by

B^--N^+ and obtain a neutral species isoelectronic with the original one. To his anxious query, "Will they be liquid?" I replied, "Who can tell? Nobody has made such a compound. Let us hope they will be."

So this project, which lasted nearly 20 years and led to a long series of publications,[39] started as an attempt to prepare new liquid heteroaromatic compounds. In this we were in fact successful. One of our early compounds turned out to be liquid at room temperature. I should add that the student in question did much excellent work in this area and accompanied me, as a postdoctoral student, to Chicago on my move from London.

These "borazaro" compounds, as we termed them, have received surprisingly little attention, given the amount of interest that has been taken in other novel aromatic systems. They show none of the chemical reactivity normally associated with organoboron compounds, behaving like normal aromatic species. Perhaps the most interesting is 10,9-borazaronaphthalene (**34**), which not only looks like naphthalene but even smells like naphthalene.[40] It also failed to react with maleic anhydride: remarkable, given that naphthalene itself forms significant amounts of the Diels–Alder adduct at equilibrium.

Phosphonitrile Chlorides

Another potentially related problem concerned the structures of the phosphonitrile chlorides. The structures are polymers of phosphonitrile chloride ($PNCl_2$), consisting of alternating PCl_2 groups and nitrogen atoms, forming either a ring or a chain with suitable terminating groups. Their structures had aroused interest because of their stability and because of the possibility that they might contain potentially aromatic cyclic conjugated systems of a novel type, involving d atomic orbitals (AOs) of the phosphorus atoms and p AOs of the nitrogen atoms. Such con-

33   **34**

jugation could account for the fact that the PN bonds in such compounds are very short. Furthermore, Craig had pointed out that the interactions between p and d AOs differ from those between p AOs in that the phases of the orbital lobes on one side of a d AO are opposite to those on the other and that this could lead to the rules for aromaticity in the phosphonitrile chlorides being different from those for cyclic polyenes.

We pointed out[41] that extended conjugation of this type is in fact impossible in the phosphonitrile chlorides because π bonds formed by d orbitals show the same type of directivity as σ bonds formed by p orbitals and because the three-atom (NPN) bond angles (~110°) are too small. A phosphorus d AO cannot therefore overlap simultaneously with the p orbitals of the two adjacent nitrogen atoms. On the other hand, phosphorus has 5 $3d$ AOs available, and it is easily seen that two of them can be chosen in such a way that each overlaps effectively with the p AO of one of the adjacent nitrogen atoms. Thus the π orbitals in the phosphonitrile chlorides are not cyclic. They consist of NPN units that interact very little with one another, leading to localized three-center π bonds. This picture, which soon became generally accepted, accounts not only for the stability of the phosphonitrile chlorides but also for the lack of any significant differences between ones with different-sized rings or between the cyclic and linear types.

The Condensation of Steam on Metal Surfaces

Another amusing project arose from a lunch-time meeting with the professor of mechanical engineering. He explained to me that the rate-determining factor in the condensation of steam on metal surfaces is the poor conduction of heat through the film of liquid water that forms on the metal. If steam can be induced to condense in drops, like mercury, it condenses 20 times faster. He asked me if I thought something could be done about this, and when I said "yes," he asked whether I would be willing to do it. I said I would if he could get support for the project, which I assumed would end the matter. However, to my surprise, he did, so I was stuck with it. Luckily, I had a good graduate student who expressed interest in it, so we got to work.

The goal was obvious. Water normally condenses in a film because it wets metals. We had to make the metal surface water-repellent. The approach to be followed was also obvious. We had to find some way to chemically coat the metal with hydrocarbon. Because condensers are made of copper alloys, the obvious anchoring group was something containing sulfur or selenium. Because I assumed that this must have been tried and found inadequate, my idea was to try out all the groups we could think of that had sulfur or selenium in them, pick the best, and then make a polymer anchored all the way along by a series of such groups.

So my student, who has ended up in the higher regions of industry in Britain, made about 80 stearyl derivatives containing sulfur or selenium.[42] We built a device to test their ability to promote dropwise condensation on copper, the object being to find out which anchoring group would survive the longest. The project was, however, *too* successful. Every single compound promoted dropwise condensation perfectly, and none of them, once on a copper surface, ever lost its effect![43] Even distearyl sulfide was still promoting perfect dropwise condensation after 4 months!

So our carefully organized plan to find the best anchoring group turned out to have been quite unnecessary. We discovered later that no one had ever tried anchoring groups of this kind. All the previous work on dropwise condensation had been carried out by chemical engineers who used easily available compounds, such as stearic acid, which were rapidly washed off the metal by the condensing steam.

We ended up testing one of our compounds in one of the condensers of a Britain–France ferry boat operating on the Southampton–le Havre route—a story in itself. When we abandoned the experiment a year later, the treated surfaces were still condensing steam in drops. I have no idea what happened subsequently to our contribution. It certainly aroused a lot of interest at the time—too much as far as I was concerned. Parties of engineers from British Electricity, the British Navy, and goodness knows who else kept invading my office with blueprints of condensers, asking how they should best be treated. Finally, I managed to dump the whole project on an amenable younger member of the faculty.

Liquid Crystals

Another project was concerned with the use of liquid crystals as solvents. At that time, the study of liquid crystals was in limbo, no further work having been done since early pioneering investigations had established their existence and general nature. It had occurred to me at Courtaulds that foreign molecules dissolved in a nematic liquid crystal should be oriented parallel to the molecules of the solvent, and that such orientation might lead to an acceleration of polymerizations conducted in liquid-crystalline solutions. Bamford and I never pursued this, and it now occurred to me that such orientation could also be useful in spectroscopy.

The first step was to see whether one could obtain reasonable concentrations of solutes in nematic liquid crystals without destroying their anisotropy. I sold this project to a member of the faculty at an Indian university who had come to QMC on leave of absence to work with me for a Ph.D. He took up the problem with great zeal, and his phase diagrams[44] for a number of such binary systems provided basic information concerning the conditions under which liquid-crystalline solutions exist.

The next step would have been to use nematic liquid crystals as solvents for rod-shaped molecules and to study absorption of polarized light by the resulting solutions. However, although my student continued work on liquid crystals on returning to India and indeed became an authority in the field, I was never able to find another student to carry on our work. Whenever I suggested it to prospective candidates, their immediate reaction was to ask where they could read up about the problem. When I told them there was nothing to read because virtually no work had been carried out on liquid crystals, they immediately assumed something had to be wrong with the whole area. So my project came to an untimely end.

Nuclear Magnetic Resonance Spectroscopy

During my time at QMC, NMR made its appearance, so I wanted to get an NMR spectrometer. Unfortunately, the only commer-

Being congratulated by Sir Lawrence Bragg, president of the Chemical Society (now the Royal Society of Chemistry), after giving the Tilden Lecture in 1954. The Chemical Society's named lectureships are its equivalent of awards and are now accompanied by a medal. When this new policy was introduced, the society gave medals to all surviving previous lecturers, so I now also have a Tilden Medal. I thought this a very nice gesture.

cial instrument at that time was one made by Varian, and that cost dollars. I might have been able to get the money to buy one in pounds. At that time, however, one had to get special permits to convert pounds into any foreign currency. Because there were already two NMR spectrometers in London, one at University College and the other at Imperial College, we would have had no chance. A bit unfair, given that the different colleges in London were, in effect, independent universities, so we had no access to equipment in any of the others.

At this time, Rex Richards, at Oxford, had developed a high-quality permanent magnet and had used it to construct an effective NMR spectrometer. One of the major electronics companies in Britain was trying to commercialize his magnet, so I ordered one. However, the students in my group were organic chemists, none of whom had ever built any equipment or knew anything about electronics. I did know a little, but not very much. So we needed to gain experience before our magnet arrived. Had I known how long it would take to come, I would not have worried. When I left London, some years later, we were still waiting. We did not know at the time that the company concerned was as incompetent as it proved to be.

Anyway, the obvious thing to do seemed to be to build a low-resolution proton NMR spectrometer for solids that would involve similar circuitry and require only a cheap magnet. I also succeeded in selling the project to a good student. However, disaster struck. Normally, the only source of support for graduate students was a state-operated system of grants. To get one of these, a student had to have a B.Sc. degree with at least upper second-class honors. The student in question had a brainstorm in his final degree examinations. Instead of getting a first or upper second, as expected, he ended up with only a lower second.

Lunch with another of our engineers, this time the professor of civil engineering, saved the situation. He was interested in concrete. He asked me if I had any ideas about the differences between good concrete and bad concrete, what they might be caused by, how they might be studied, and whether I would be willing to carry out such a study. The answer was, of course, obvious: solid-state NMR! And yes, but only if he could find money for the project. He was delighted by the idea, and at his suggestion and with his support, I sent a proposal to the appropriate national laboratory. It got glowing reviews, the project was funded, and my student got his grant and duly built the spectrometer.

As his main interest was building it, he was quite happy to study concrete with it. Dozens of samples came in: good concrete, bad concrete, every kind of concrete. All of them gave identical NMR spectra. Our only positive result was a method for studying the setting of concrete. Because ice gives no NMR signal, we could follow the disappearance of water while concrete was setting by freezing samples and observing the growth of the proton signal in them. This work was enough to get the student his Ph.D.

I did not, of course, want to continue work on concrete. My final coup was a report in which I explained that our work had proved disappointing, apart from our method for studying kinetics, and that I did not feel justified in asking for the grant to be continued. Because research money was hard to come by, I ended with enthusiastic praise from the people who had given me the grant for my public-spirited attitude.

Electron Spin Resonance and Nuclear Quadrupole Resonance

Although our high-resolution magnet failed to materialize, this work had two sequels. First, we used the magnet from the NMR spectrometer to build a very effective electron spin resonance (ESR) spectrometer, by using war surplus radar equipment for the microwave circuitry. Second, we carried out some of the early work on the then-novel area of nuclear quadrupole resonance (NQR) spectroscopy, again building our own spectrometer.[45] Building this spectrometer proved a bigger undertaking than we had expected, because the chlorine NQR band was saturated at one end by radio amateur transmitters and at the other by the London TV station. Enclosing the spectrometer in a copper box inside another copper box failed to deal with the interference. We finally dealt with it by building a small screened room in an old hut outside the chemistry building, a war relic, using, for economy, wire netting of the kind used to make chicken coops. Even this proved inadequate until we spot-welded every single join in the mesh. I have long felt that anyone with a good training in organic chemistry is ideally qualified for work in any other area of science. Our work on magnetic resonance spectroscopy is just one of my many pieces of evidence. When we began, none of my students knew anything about electronics, and my knowledge was strictly nonprofessional. By the end, students working on magnetic resonance in the physics department were coming to ask my students for advice.

Projects for Graduate Students

While at QMC, I developed the pattern I have followed in all my subsequent work. Most of this has been carried out by graduate students. I have always had very strong views concerning the kind of problem a Ph.D. student should be given. The purpose of the Ph.D. course is, or should be, not to carry out research of scientific interest, but rather to teach students how to carry out research. Projects for graduate students should be chosen on this basis. Each student should have an individual project of his or her own and it must be one in which he or she is interested and that he or she can be expected to finish in a reasonable length of time, that is, the normal time expected for a Ph.D. course.

I have always objected very strongly to the use of graduate students as cheap slave labor. This would have been unthinkable at Oxford when I was there, and it is a pity that the same is not true everywhere today. To me it seems immoral to give students speculative projects or to use them as a part of a research team. Projects of that kind should be reserved for postdoctoral students. Because I had no postdoctoral students at QMC, other than Roly Pettit, my research was correspondingly limited. The only large projects related to areas of chemistry. Each of the students I had working in such an area had a clearly defined and self-contained project chosen by himself or herself from a varied list. So varied was this list, that on looking through my list of publications while preparing this article, I found a number describing projects I had completely forgotten. If anyone is interested, write and ask for a list of my publications!

This work was, of course, largely dictated by the facilities available at QMC, which, at first, were almost nonexistent. Even when I left, they were still barely adequate. I have also always been plagued by my inability to persuade people to do things. When the success of a given project has suggested an obvious follow-up, I have often been unable to find anyone willing to do the job. My inability to sell pet projects to graduate students need not, of course, have mattered in London. Professors in Britain had dictatorial powers. As the professor of chemistry, I could simply have told my students what to do. Most professors did just that. Indeed, because the professor had complete control over his department's budget, he could also pressure other members of his faculty to work with him, by making it clear that that was the only way they could hope for research support. Many professors did just that. At University College, for example, all the organic faculty found it expedient to collaborate with Ingold. I was unusual in letting graduate students pick their own supervisors, let alone their research problems.

I only once took advantage of my authority. A very good graduate student, now in a high administrative position in Eastman Kodak in Rochester, set off on a project in which I was particularly interested. It would in fact have led to one of the first demonstrations of oxidation by electron transfer in an organic reaction. However, after only 6 months, he became discouraged and asked to change to something else. I was incensed.

Now, at that time, one of my particular hates was the then-current drive to distinguish between different substituent effects on the basis of the Hammett equation, by using data for benzene derivatives only. Because there are just two pieces of data for each substituent (i.e., σ_m and σ_p), attempts to partition these into contributions by two different substituent effects, resonance and inductive, seemed to me utterly futile. Whether attempts of this kind are of any real value is a matter of opinion. However, if they are going to be carried out at all, they should at least be carried out properly. This meant finding some source of additional independent data, which in turn meant studying substituent effects in some system other than benzene. Such a study would, of course, involve a lot of arduous and not very rewarding experimental work. Just the thing for a recalcitrant graduate student!

I told him that if he wanted to change, he would have to make a series of derivatives of α-naphthoic acid and of β-naphthoic acid, with eight different substituents in each of the five unhindered positions in each acid, and measure their pK_As. This would increase the number of independent pieces of data for each substituent from 2 to 12 and so allow sensible deductions to be drawn. Quite an assignment, given that synthetic procedures then were much more primitive than they are today! However, he rose to the challenge.

Although he did not make all 80 compounds, he made most of them, and enough, in each case, for estimation of σ constants—a heroic effort. We were, moreover, able to show that these σ constants could be estimated with reasonable accuracy by a simple PMO treatment in terms of resonance and field effects, without reference to the classical inductive effect.[46] Thus one could get quite reasonable estimates for the σ constants of substituents at any position in any AH, and for any position of the reaction center, by using just two parameters for each substituent that corresponded to its mesomeric (resonance) and field effects. We, therefore, called this approach the field-mesomeric (FM) method.

At that time, all studies in this area were based on the assumption that inductive effects are significant and field effects unimportant. Our work indicated that the classical inductive effect is, in fact, unimportant, a conclusion that incidentally follows from quantum mechanical calculations. The polarity of the bond linking a substituent (S) to the adjacent carbon atom contributes to the electrostatic field generated by the substituent. It does not lead to significant polarization of adjacent CC bonds, as had been, and still is, commonly supposed. The polarity of the CS bond does, however, lead to a change in electronegativity of the carbon atom that can lead to polarization of the π electrons of a conjugated system involving that carbon atom. The resulting charges set up electrostatic fields that can then influence the rate of reaction by corresponding field effects.

This is the so-called π inductive effect. Later, we extended the FM treatment to take account of this, leading to a treatment that we termed the field-mesomeric–mesomeric-field (FMMF) method. Table I shows the results for benzene and naphthalene derivatives.

Roads Not Followed

Our work should have brought rationality into the study of substituent effects. It was in fact quietly forgotten because I never followed it up. This indeed illustrates a general problem that many others have also encountered. Those working in a

Table I. Calculated σ Constants at Various Ring Positions for Naphthalene and Biphenyl Compared with Experimental Values Derived from α- and β-Naphthoic Acids and Biphenyl-4-carboxylic Acid

Substituent	3 (3')	4 (4')	5	6	7
			α-Naphthoic acid		
CH_3	−0.06 (−0.05)	−0.23 (−0.14)	−0.11 (0.01)	−0.06 (−0.05)	−0.11 (−0.07)
Cl	0.38 (0.30)	0.15 (0.26)	0.21 (0.29)	0.28 (0.17)	
Br	0.40 (0.34)	0.16 (0.30)	0.23 (0.30)	0.29 (0.18)	0.21 (0.07)
OH	0.13 (0.06)	−0.59 (−0.52)	−0.11 (−0.06)	0.08 (−0.08)	−0.15 (−0.10)
OCH_3		−0.47 (−0.36)	−0.10 (−0.01)	0.07 (−0.06)	−0.13 (−0.08)
CN	0.55 (0.59)	0.73 (0.79)	0.50 (0.46)	0.44 (0.34)	0.48 (0.31)
NO_2	0.70 (0.61)	0.84 (0.86)	0.61 (0.54)	0.55 (0.41)	0.59 (0.36)
			β-Naphthoic acid		
CH_3	−0.09 (−0.09)		−0.09 (−0.05)	−0.05 (−0.05)	−0.09 (−0.07)
F			0.05 (0.07)	0.15 (0.14)	0.03 (0.08)
Cl	0.35 (0.26)		0.13 (0.16)	0.18 (0.18)	0.04 (0.06)
Br	0.38 (0.25)	0.17 (0.18)	0.14 (0.17)	0.19 (0.19)	0.05 (0.06)
I	0.34 (0.22)		0.11 (0.15)	0.17 (0.18)	0.03 (0.05)
OH	0.09 (−0.11)	0.02 (−0.03)		0.03 (−0.09)	−0.18 (−0.14)
OCH_3	0.07 (−0.01)	−0.01 (−0.01)	−0.13 (−0.11)	0.01 (−0.01)	−0.18 (−0.01)
CN	0.58 (0.57)	0.30 (0.37)	0.37 (0.35)	0.31 (0.35)	0.26 (0.24)
NO_2	0.73 (0.60)	0.38 (0.40)	0.44 (0.45)	0.40 (0.37)	0.30 (0.28)
			Biphenyl-4-carboxylic acid		
CH_3		−0.06 (−0.02)			
Cl		0.09 (0.13)			
Br	0.13 (0.12)	0.10 (0.13)			
OH		−0.09 (−0.19)			
OCH_3		−0.08 (−0.07)			
NH_2		−0.22 (−0.25)			
NO_2	0.25 (0.23)	0.31 (0.30)			

given restricted area of chemistry usually form a cozy circle of self-styled experts with mutually accepted dogmas. Any real innovation has to come from outside, and any such intrusion is strongly resisted by the members of the circle. Unless the out-side contribution is followed up, they quietly forget it, and because they are regarded outside as the "experts" in the area in question, the new contribution is generally ignored. The excuse, in this case, was that our procedure did not lead to "accurate" results. Because correlations of the kind illustrated by the Ham-mett equation are not expected to be quantitative, this was hardly a valid criticism.

This compartmentalization of chemistry has also had the unfortunate effect of making publication of unorthodox papers difficult. Papers in a given area are naturally sent to the "experts" in the area as referees. Any paper that fails to toe the party line naturally runs into trouble, particularly if the author is not a member of the relevant clique. Getting such papers into print can take a lot of effort and wasted time. The unfortunate thing is that really novel ideas in *any* area tend to come from outside, precisely because of the pressure for conformity on the part of the insiders.

Another of my QMC projects provides a good example of this problem. A graduate student, who came from Canada, wanted to work with me on some problem related to solid-state reactions. At that time, very little was known about the course of chemical reactions in the solid state. We decided to study the thermal decompositions of some carboxylic acid derivatives. The work turned out well, and we were able to draw conclusions concerning the mechanisms involved. This work was, however, never published, other than in the form of a Ph.D. thesis.[47]

Work in the area at that time centered entirely on the physics of thermal decompositions, not their chemistry. The interest was entirely in the role of crystal defects, the propaga-tion of dislocations, etc. When I submitted a paper reporting our work to *Transactions of the Faraday Society*, the referees attacked it violently on the grounds that it showed complete ignorance of the way research in the area should be conducted. Because I had a lot of things on my hands at the time, I aban-doned the paper. Today, 30 years later, mechanistic studies of solid-state reactions have become fashionable. Even now, much of our work would probably still be of interest.

A lot of my work has been lost in this way. Having been mostly self-taught, so far as chemistry is concerned, I have always tended to be something of a loner, working on my own and without much concern for "establishment" views. For the same reason, my work has covered a very wide area of chemistry. A lot of my projects have also been "one-off" jobs, having been based on two further principles acquired from Collingwood: first, that research is most effective when it is designed to answer specific questions, and second, that although no amount of positive evidence can prove a theory, *one* piece of negative evidence is enough to disprove it. If an experiment gives an unambiguous answer to the question it was designed to answer, there is no need to carry out further work in that particular connection. Likewise, if an experiment disproves the theory it was designed to test, further tests are unnecessary.

In practice, however, a single paper reporting such an experiment usually has little effect. People understandably tend to judge the significance of a piece of work by the number of publications it generates. It is therefore not enough to have a new idea or to disprove an old one, particularly if the old one is well-entrenched. One must be prepared to spend time and effort on publicizing it. I have always been very bad at publicizing the things I have done. The main motive behind my research has always been the desire to find out something I wanted to know. Having found it out, I go on to something else. To spend more time on a problem I have solved to my own satisfaction, simply to convince others, has always seemed to me a waste of time. This is not a commendable attitude, by any means. It is plain self-gratification.

The worst example of such self-gratification is PMO theory. Having produced a complete theory of organic chemistry in the six original papers in the *Journal of the American Chemical Society*, I saw no reason for following these up with applications to specific problems. I should, of course, have written a book on PMO theory. Such a book would have formed a very effective sequel to my first one. Because I did not do so, very few chemists today realize the theory's potential. When Ralph

Dougherty and I finally published an account of it in book form,[48] in 1975, it was too late. How could anyone be expected to believe that a theory more than 20 years old, which had not been mentioned to them by their mentors or described in any of the standard texts, could possibly be of any significance? The curious thing is that very little had to be added to take account of the work done in the intervening years, and very little more needs to be added today. Perhaps Ralph and I should have presented it as something entirely new, without any reference to the earlier work!

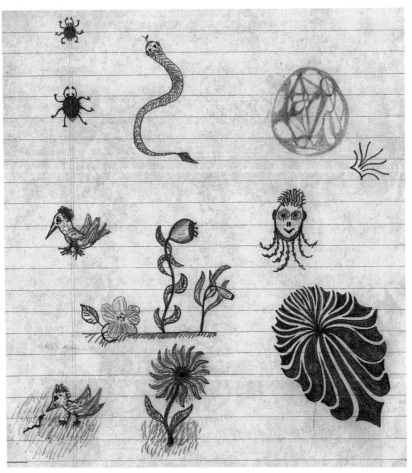

Notes from a particularly awful meeting of some London University committee.

Mary and I at a Chemical Society reception in 1958.

To be fair, I did have an excuse. The main reason I did not write such a book is that I simply could not find the time to write it. Apart from my problems in the Chemistry Department at QMC, and apart from the various committees I had to attend as head of the Chemistry Department, and apart from the Board of Studies in Chemistry (a vast central committee, composed of chemistry faculty from all the colleges in London and responsible for supervising courses, examinations, etc.), I also kept getting appointed to other major committees of the University of London. I ended up on five of them.

Being appointed to such a Committee was supposed to be an honor. To me, any committee is a complete waste of time. Either the other members of a committee think the same way I do, in which case I might just as well not be there, or they disagree, in which case I have to waste time arguing with a bunch of obstinate idiots. Furthermore, the meetings were held in the Senate House, in the middle of London, and QMC was in the East End, so half my time was spent on traveling to and fro.

The last straw was my election as one of the two Honorary Secretaries of the Chemical Society (now the Royal Society of Chemistry), a 6-year appointment, and a very important one that involved a lot of work. The Chemical Society was in yet a third part of London. At this point, the situation had become totally desperate. I had no time at all for research, and most of my life was being spent in London traffic jams, trying to get from one meeting to another. There was only one way out: to leave Britain. At this point, the University of Chicago offered me the professorship that had become vacant with Kharasch's death—so we went.

The Move to Chicago (1959–1963)

We were, in fact, well-prepared for the move. Three years earlier I had finally, by various subterfuges, persuaded QMC to appoint a second professor of chemistry, a very unusual move at that time in Britain. I immediately seized the chance to accept an invitation from Yale to spend a semester there as visiting professor. Mary and I went for the spring semester in 1957 and stayed in America for 6 months. It was not only a wonderful experience but also a very happy one. We loved America from the start.

We also saw most of it. We bought an old Dodge in New Haven, and after leaving Yale, set off in it on a vast tour round the country, seeing everything and visiting all the major Universities. Our friends at Yale were appalled at the thought of our making such a trip, let alone in an aged car. We were presented with numerous water bottles to save our lives in the desert. The thought of us dying of thirst clearly filled everyone with horror. The only really sensible contribution came on our last evening, when Bill Doering arrived at the door of our rented house with a case of Scotch, remarking that no one had ever been found dead in the desert with a full bottle. We took him at his word. When we reckoned up at the end of the trip, Mary and I found we had done 5000 miles per gallon.

This trip also let our sons, Robert (who was then 12) and Steuart (who was then 10) visit America. They were then at

W. von E. Doering at Yale University. Bill is one of our oldest and closest American friends. He was, I suspect, responsible for my invitation to Yale as a visiting professor in 1957, and he energized our subsequent tour of the country by a well-chosen leaving present.

Copthorne, the preparatory school in England that I had been to as a boy, and flew across for the Easter holiday. As I have already mentioned, the school year in Britain consists of three terms, each lasting 10–12 weeks, with a month's holiday at Christmas, a month in the spring, and 2 months in the summer. So Robert and Steuart were able to spend a month with us, giving us time to show them a lot of America. This travel was a big help to us a year later, when we were thinking of moving there permanently, because our main concern over moving was the effect the move might have on them, particularly Robert, who had just moved to St. Paul's, a leading public school in England, where he had won a scholarship. The fact that they both had had a euphoric time in America 2 years earlier was a big help. Even so, things were tricky because they were both very intelligent and they were both far ahead of their contemporaries at American schools because English schools have a more concentrated emphasis on academic work in the early years of education. Fortunately, the University of Chicago's Laboratory School was exceptional in allowing students to proceed at their own pace in each subject. Furthermore, those who progressed beyond the school level could take courses at the university itself. Robert had indeed taken the first 2 years' courses in

Mary, Robert, and Steuart in London in the late 1950s.

chemistry at the University of Chicago when he went there at the age of 16 as an early entrant, and Steuart graduated normally from school when he was 15. Indeed, their main problem turned out to be the shock of meeting girls for the first time, private schools in England being then almost universally segregated. The move certainly turned out well for both of them.

My Sons

Robert majored in chemistry at Chicago when he was 18. He stayed there for his Ph.D. because by the time he got his B.S., he had already taken all the graduate courses in chemistry. He had intended to become a chemist and indeed spent a year working on an inorganic project with Jack Halpern. However, at this point he serendipitously developed a passionate interest in computers and switched to working with Everly Fleischer, writing computer programs for X-ray crystallography. When he got his Ph.D., the computer scientists at Chicago wanted to appoint him an assistant professor, but the dean refused on the grounds that he had no formal qualifications of any kind in computer science. So he went instead to the Illinois Institute of Technology (IIT), which was less fussy. Some years later he moved to NYU as professor of computer science at the Courant Mathematical Institute, where he is now a leading authority on high-level computer languages. He got a presidential citation for writing the first certified compiler for ADA. ADA is a new computer language that the Department of Defense (DOD) commissioned. All DOD programs now have to be written in it.

Steuart has also had a very successful career in computers. He currently owns a company in Chicago (DISC: Dewar Information Systems Corporation) that deals in computer editing systems for small and medium-sized daily papers, over 25% of which are now using DISC systems. He now has more than 100 people working for him.

Steuart also got into computing by accident. He originally went to Columbia to major in biochemistry but decided after a year to return to Chicago to major in music. He did well, being within one course of getting an M.A. at the time he got his B.A., and being strongly urged by the music faculty to try for a career as a concert pianist. At this point, however, the draft caught up with him. Steuart was not prepared to fight in Vietnam because he, and we, regarded that particular war as wholly immoral, given that its purpose was to prevent free democratic elections from being held in Vietnam. Steuart could have avoided the draft quite legally by leaving the country because he was then still a British citizen. However, because he wanted to stay in the United States, he decided to apply for conscientious objector status. We thought his chances slim because he had to face a Texas draft board (by then we had moved to Texas), and, like both of us, he was not even a member of a church. However, we were wrong. Steuart not only passed but clearly impressed the board greatly. Indeed, the chairman's parting words were, "We wouldn't like you to think of us as warmongers, you know"! In place of military service, Steuart was required to work for 3 years at Billings Hospital, the

Robert and Steuart with the 8-in. telescope they built in the workshop at the Adler Planetarium in Chicago.

University of Chicago's medical school, on the pay he would have had in the army, the assumption being that he would be given menial jobs that needed no training. However, when the hospital authorities heard that he had spent a year at Columbia University as a biochemistry major, they put him in the pathology laboratory, running tests, and when the man in charge of the hospital computer decided to leave, Steuart got Robert to teach him how to program the computer and applied for, and got, the job. He did very well at it, becoming quite an authority on the use of computers in medicine. Indeed, he presented papers at several conferences. At the end of his 3-year stint, the hospital offered Steuart a permanent appointment at quite a high salary. However, he decided instead to join a Chicago computer company. He left it a few years later to found DISC.

So both Robert and Steuart started in chemistry but later moved to other fields. I was greatly relieved that they did. It is rarely a good thing for sons to try to follow in their fathers' footsteps, especially in chemistry.

My move to Chicago shook up the British establishment. The following spring I was elected a fellow of the Royal Society, and various attempts were made to get me back. However, it was too late. Although the "brain drain" had been in operation for some years, I was the first professor to defect. Indeed, I remained the only one for many years. Being a professor in Britain had a lot of advantages. Salaries were still reasonable then, and professors enjoyed complete independence, high social status, and dictatorial powers. The last two meant less to me than they did to others.

One person who benefited from my move was Derek Barton, who was then the professor of organic chemistry at Imperial College (London University). Derek had been trying, without success, to get something done about the laboratories at Imperial College, which were relics from the past and wholly inadequate. After I moved to Chicago, Derek also received an approach from across the Atlantic. When news got to the British Cabinet that he was seriously considering the offer, they were horrified at the thought of a second defection and immediately got in touch with Imperial College. I have a note from Derek with a postscript saying, "You will be amused to hear that I am getting my new building after all!"

We very much enjoyed Chicago. Our 4 years there were very happy ones. We felt at home in America, a feeling that has become even stronger with time. We made a lot of friends, and for the first time in many years I was able to concentrate exclusively on chemistry. My only involvement with administration came early on. I was put on the building committee, which at that time, was mainly concerned with plans to modernize the Kent Laboratories, a chemical mausoleum on a par with QMC. My immediate reaction was "why not build a new chemistry laboratory?" No one had thought of this. Everyone agreed that it was a wonderful idea. So having planted the seed, I hastily resigned from the building committee, and left it to others more qualified to put the plan into operation.

I can thus claim credit for having initiated Chicago's magnificent new chemistry building. It was, of course, built after I left. I seem to have spent my whole career working in chemical slums, in places that were about to get new chemistry buildings but did so only after I had moved elsewhere. The one exception was Austin. However, the new building there (Welch Hall) was so badly designed that I don't think it ever really counted as a new building.

Chicago Projects

My main project in Chicago was a continuation of our work on borazaromatic compounds and involved synthesis of a number of additional examples and studies of their chemistry. This was good classical stuff, but it did not lead to any very exciting conclusions. Indeed, the really exciting thing about borazaromatic compounds was that they turned out not to be in any way remarkable. They behave just like other benzenoid heteroaromatic compounds, being quite unlike normal organoboron compounds. They are stable in air, undergo typical aromatic reactions such as electrophilic substitution, and their

How to make morphine? D. H. R. Barton with me at the Ciba symposium in honor of Sir Robert Robinson. I got Derek a new chemistry building at Imperial College.

spectroscopic properties are, again, just what would be expected. Much of this work is reviewed in ref. 40.

One compound, which we studied as a sideline, did, however, have some entertaining aspects. Treatment of *o*-aminomethylbenzeneboronic acid with formaldehyde had been claimed to give a dimeric species (**35**, $n = 2$) with an eight-membered ring. This structure seemed unlikely because it would clearly be very strained. Reexamination[49] suggested that it was, in fact, **36** with a kind of internal sandwich structure because analysis indicated it to correspond to a monohydrate of the trimer (**35**; $n = 3$). The water was held very strongly; yet IR spectroscopy indicated no hydrogen bonding.

Anhydrous crystals of **35** could be obtained by crystallization from anhydrous solvents, and the trimeric structure was confirmed by the size of the unit cell, determined by X-ray crystallography. Models indicated that the water could be held as a clathrate in the space between two adjacent molecules of **36**, in a

35

36

region where there was nothing with which it could form a hydrogen bond. The only way to confirm this intriguing structure was by X-ray crystallography. The unit cell was, however, enormous, containing four molecules of **36**. It also lacked symmetry, and **36** contains no heavy atoms. The structure was, therefore, insoluble by existing X-ray techniques.

However, our older son, Robert, was then working for his Ph.D. in chemistry, at Chicago, with Everly Fleischer, developing computer programs for X-ray structure determination by the direct method. At that time, the method had not yet been extended to crystals with unsymmetrical unit cells. Robert wrote the first program to do this. Obviously, he should have tried it out first on something simple. However, because Everly had the data for our compound, Robert thought it would be fun to try his program on it—and it worked! Our predictions were correct, so our compound was the first with an unsymmetrical unit cell to have its structure determined by the direct method.

For reasons I have already discussed, most of my work in Chicago, as in London, was concerned with "one-off" projects, each designed to answer some specific question. Projects of this type have the additional advantage of providing ideal training for graduate students. One, for example, was concerned with a long-standing problem in π-complex chemistry: the relative energies of a proton π complex (**37**) and an isomeric carbenium ion (**38**). Studies[50] of the addition of deuterium bromide or hydrogen bromide to acenaphthalene, and to 1-phenylpropene, showed that *cis* addition predominates. Thus the first step, in

$$\underset{\mathbf{37}}{\overset{\overset{\displaystyle H^{+}}{\uparrow}}{\diagdown C = C \diagup}} \qquad\qquad \underset{\mathbf{38}}{\overset{\displaystyle H}{\diagdown} \underset{\diagup}{C} - \overset{+}{\underset{\diagup}{C}}}$$

solution at least, involves the formation of a classical carbenium ion, not a π complex.

Another project[51] involved the preparation of metal coordination polymers from transition metals with bifunctional aromatic ligands as potential semiconductors, following a theoretical argument that such species might have high carrier mobilities. While the compounds we made were not really suitable, our main purpose was to find an excuse for publishing the theory. I still think this idea promising. Various vicissitudes have prevented me from pursuing it as effectively as I would have liked.

Another project was concerned with the structures of molecular complexes, in particular those formed by aromatic molecules, (e.g., picrates of aromatic hydrocarbons). I felt strongly that Mulliken's charge-transfer theory of the bonding in these compounds was incorrect. As Briegleb had pointed out many years earlier,[52a] the heats of formation of such complexes can be explained in terms of normal van der Waals forces. There is no need to look further. We set out to test this suggestion. If it was correct, the spectrum of such a complex should contain three sets of bands: two corresponding to local excitations of the individual components and differing little from those of the individual components, and the third to charge-transfer transitions due to transitions from a filled MO of one component to an empty MO of the other.

Although local excitations had been observed, they had been attributed to dissociation of the complexes in solution. We avoided this difficulty by measuring the spectra of the solid complexes in potassium bromide discs. The spectra still showed the locally excited bands of both components, shifted little from those for the separate components.[53] Furthermore, if such a complex does indeed contain essentially unperturbed molecules of the components, the energy of transition (ET) of the first charge-transfer transition should be equal to the ionization energy of the donor *minus* the electron affinity of the acceptor, *minus* the electrostatic energy of interaction between the resulting ions. In complexes formed by various "donors" with a given "acceptor", ET should be equal to a constant plus the ionization energy of the donor. This proved to be the case, and when more than one charge-transfer band was observed, the difference between the two ETs could be related to different ionizations of the donor.

It is amusing to note that in the second version of his book,[52b] Briegleb adopted the Mulliken interpretation. He should have stuck to his guns! Our later studies, which will be discussed presently, have shown fairly unambiguously that charge transfer plays only a minor role in the binding of such complexes.

The Start of My Theoretical Program

However, the most important new development was the start of my current theoretical program, which began almost by accident, as a result of an idea I had had at QMC. It occurred to me that the repulsion between two electrons occupying a p AO should tend to keep them in opposite lobes of the AO, and that this type of correlation could be taken into account by using lobes of p AOs as independent orbital functions. This idea was studied by an Argentinian student whose English, although almost perfect, sometimes missed subtle points. She therefore, quite naturally called the lobes "*split-p orbitals*[54]," a splendid term that I adopted with great joy.

The resulting split-p orbital (SPO) approximation caused something of a furor. I described it at one of Charles Coulson's theoretical conferences in Oxford, where it roused Boys to a passionate outburst—the only time I ever saw him in such a state of agitation. Indeed, he was so upset that he retired to bed with "flu" and missed the rest of the conference. I am sure it was just a psychosomatic effect of theoretical shock. The split-p orbital method must have caused more lost sleep than most heretical ideas. Everyone agreed that it was ridiculous, but no one could see why. Finally, Griffiths came up with the one real objection: because individual lobes of AOs are not orthogonal

Professors A. and B. Pullman. Alberte and Bernard were among the early practitioners of MO theory.

to the core AOs, the SPO approximation is inconsistent with the separation of σ and π electrons. However, the idea in itself has survived in the form of the Gaussian lobe orbital approximation in ab initio theory.

Although the SPO method, itself, fizzled out, it got me interested in the possibilities of MO calculations, not that I thought that anything much was likely to come of them. At that time, it seemed extremely unlikely that MO calculations would ever lead to results accurate enough to be of any real chemical value. Besides, carrying out calculations at Chicago was very difficult because the university charged for use of the computer (then an IBM 7090) at full commercial rates.

The thing that really got me started was an invitation from Bell Telephone Laboratories in Murray Hill, at the instigation of Ed Wasserman, to spend a month there in the summer, giving a few seminars and otherwise carrying on with my own work. This visit became an annual event for many years and introduced me to computing under the tutelage of Larry Snyder. At first, my computations were confined to Hückel calculations,

Meeting at Oxford, 1961, organized by Charles Coulson. Most of the leading quantum chemists and theoretical chemists were present. For a more complete list, see the next page. Courtesy Gillman & Soame, Oxford.

but later they were expanded to π-SCF (self-consistent field) calculations using the Pople π-SCF approximation for which Larry wrote a program.

Furthermore, Bell let my group use their computer by remote access (which, in those days, meant mailing boxes of punched cards) during the rest of the year. This arrangement led us to try parameterizing the Pople method to reproduce ground-state energies, just to see how well it would perform. To my surprise, it performed not only well but unreasonably well.[55] As a result, I changed overnight from being convinced that theoretical calculations could never serve any real purpose in chemistry to being equally convinced that all the problems of chemistry could be solved by a similar semiempirical approach in which all the valence electrons are included. I must confess that had I known what I was letting myself in for, I might never have embarked on the project.

←

Row 1: 1st, Walsh; 2nd, Al Matsen; 3rd, Dewar; 4th, Ed Wasserman; 5th, Andrew Liehr; 6th, John Pople; 7th, Christopher Longuet-Higgins; 8th, Robert Mulliken; 9th, M. Kotani; 11th, LeSevre; and 15th, Maurice Pryce. Row 2: 1st, M. Simonetta; 2nd, W. I. Simpson; 3rd, Edgar Heilbronner; 6th, Harden McConnell; 7th, G. G. Hall; 8th, Roy McWeeny; 10th, George Porter; 12th, Joshua Jortner; 13th, Donald McClure; and 14th, J. van der Waals. Row 3: 1st, Walmsley; 2nd, R. K. Nesbet; 3rd, Whissen; 5th, Klaus Reudenberg; 6th, Leslie Orgel; 8th, W. Kolos; 9th, R. Daudel; 10th, Alan Carrington; 11th, Jorgenson; 12th, Clyde Hutchison; and 13th, G. Hertzberg. Row 4: 3rd, Charles Coulson; 5th, Herbert Strauss; 6th, John Murrell; 7th, Jack Linnett; 8th, John Platt; 9th, Don Ramsey; 10th, Thorson; and 12th, Hurley. (Thanks to four of the scientists at this meeting for assistance in identifying people in the photograph.)

The University of Texas at Austin (1963–1990)

I had no intention of leaving Chicago. I had, indeed, rejected one or two approaches when one turned up that looked unusual and interesting. This was from The University of Texas at Austin (UT), where the Robert A. Welch Foundation had just endowed a research chair. I had heard from friends in Austin, in particular Al Matsen and Roly Pettit, that the State of Texas had just developed a great enthusiasm for higher education and wanted to make the university one of the best in the country. It sounded a very rewarding project to take part in, and I also felt sure it could be done, given Texas money and with Norman Hackerman in charge.

I had met Norman some years earlier, on our 1957 trip, when he was chairman of the chemistry department. We made a detour to Austin so that I could tell him about Roly Pettit, who was being considered for a faculty position. Norman was the most impressive and able administrator I have ever come across—and self-analysis of my own monumental incapacity in that area has made me a very good judge. We had also visited Austin once or twice while we were at Chicago and thought it a very pleasant city. Because both our sons were now at college, we were free to move. After much cogitation we did.

This move again created something of a sensation. For someone to move from a senior faculty position at a university recognized as one of the top ones in the country to an unknown backwater, as UT then was, was unheard of, and, of course, the uniqueness of my move from London to Chicago added to the sensation. My arrival in Texas put the place on the academic map. Every chemist in America wanted to come and look at UT and it was on the schedule of all those visiting America from Europe. We had a steady stream of house guests. Indeed, the London *Sunday Times* even had a major article about UT in its magazine section, something it had not done before, and has not, I am told, done since for any other American university. Certainly, no other chemist could have given the university more publicity!

Whether the move was sensible from my own point of view is another matter. I gained nothing financially by it, and I lost heavily professionally. Also, although the university progressed rapidly for my first 8 years, while Norman Hackerman was president, after he left it fell apart. His successor was, unfortunately, a disaster, and when he was replaced after 4 grim years, his replacement achieved the apparently impossible by being even worse. UT slid rapidly back downhill. Almost all our personal friends in Austin, 14 of them, all senior members of the faculty, left when Norman did, and we should have left too. However, out of Scottish obstinacy, I stuck it out, in the hope that things would eventually come right. It was a bad decision.

The events at UT show in a dramatic way how much a university can gain from having an outstanding president and how quickly it can go downhill under an incompetent one. Apart from being a superb administrator, Norman also had all the right ideas about the way a university should be run. He frequently remarked that the purpose of the administration was to serve the faculty and that his job, as the head of the administration, was to see that everyone in it did serve the faculty. During his time at UT, no administrative department had more than the bare minimum of staff that it needed. Furthermore, Norman used to pride himself on the fact that if any member of

Norman Hackerman, one of the greatest American university presidents.

the faculty ran into an intractable bureaucratic problem, he or she could refer the matter to Norman personally by telephone. The only requirement was that one stated the problem clearly and in the minimum number of words. Norman in return always gave an immediate decision. He claimed, rightly, that he made no more mistakes this way than he would have by beating about the bush, and it saved time. As can be imagined, the effect on the administration was dramatic. If one got involved in a hassle through someone in the administration behaving unreasonably or bureaucratically, one could settle the matter by threatening to refer it to Norman. The errant administrator knew that one would, that Norman would listen, and that if Norman thought the administrator was in the wrong, he would tear strips off him or her for wasting everybody's time.

I should add that Norman held a joint appointment as president and professor of chemistry and that he took the minor half of this seriously. Being president of any university is a full-time job, and, in Norman's case, things were made worse by time spent with the state legislature and traveling to and from Washington. Yet Norman taught a freshman chemistry course (at 8 a.m. to get it out of the way before the day's work started), and he rarely missed chemistry faculty meetings or parties for visiting lecturers. He also had an active research group and an office in the chemistry building to which he slipped away whenever he could. And, as a grand finale, he edited a major chemical journal. Norman is a remarkable man. He was the main reason I went to UT.

His successor, Spurr, was unfortunately like a mirror image of Norman, with everything backward. He not only failed to keep bureaucracy in check but actively encouraged it. The one thing he did well was generating red tape and new procedures. The various administrative departments naturally expanded like detonating bombs; their efficiency decreased as fast as they grew, and the faculty rapidly became regarded as a tiresome inconvenience. Under Spurr's even less competent successor, the slide downhill accelerated. Worse still, the state legislature began meddling in UT's affairs. Norman had kept them in check because they respected him and knew they could trust him. Nobody could have respected either of his successors. Even worse, the state board of control seized the chance to get its fingers into the pie, seeing good pickings in a state institution with a budget not far off a quarter of a billion dollars a year.

In my case the consequences were disastrous. One of Norman's contributions to the upgrading of UT had been the realization that any leading university has to have some area where it is *the* leading one, and that the only area in which this could be achieved at reasonable cost was in computing. So UT bought a new computer that had just been developed, the CDC6600, which was at that time the most powerful in the world. Because UT got the third one to be built and because

the first two were lemons, UT had at that time the best computing facilities of any institution in the world. Norman also stated
clearly that he intended to keep UT at the top by buying
another CDC6600 if the first one became overloaded and by
buying better computers if and when they became available.
Because the only practical advantage I had gained from my
move to UT had been free access to the university computer
(then a CDC1604), this policy was a real bonanza.

Unfortunately, Norman's successors abandoned Norman's
enlightened policy. Other better computers appeared without
UT getting one, and when the CDC6600 became overloaded, it
was not replaced. At this point, the first really powerful mini-
computer, the VAX 11/780, appeared on the market, and I succeeded in getting a grant from the National Science Foundation
(NSF), with matching funds from UT to buy one. When the
grant came through, I tried to order it. At this point, the state
board of control intervened. By state law, the computer had to
be put out for bids. There was in fact no equivalent of the VAX
at that time, so my specifications made it impossible for any
other company to put in a competitive bid. Other comparable
computers were far more expensive. The state board of control
enabled another manufacturer to bid by canceling my specifications on the grounds that they were "too restrictive". The computer they were trying to make me buy was naturally cheaper
because it was grossly inferior. I also happened to know that it
would have been useless for our purposes because a colleague in
Germany had discovered that by bitter experience. The so-called
"expert" at the state board of control refused to listen to my evidence, and the president we then had at UT refused to help.
After 6 months, I finally wrote to NSF, saying that if UT insisted
on my buying the substitute computer, I would recommend that
the grant be canceled because I was not willing to be responsible
for misuse of federal funds. At this point the state board of control capitulated, and the VAX was at last ordered.

In the meantime, however, word about the VAX had got
around so that there was a large backlog of orders. Another 6
months passed before my computer was finally delivered, and it
took us many months more to get all our programs running on
it. Converting large CDC programs to the VAX is a major operation, and my group was handicapped by the fact that we were

all chemists. Furthermore, just at the time the VAX was ordered, the university's CDC6600 became so overloaded that we could no longer use it at all. This was the final blow. By the time we were once more able to carry out calculations, several key members of my group had left. My theoretical program, which now included the majority of my group, was paralyzed for more than 2 years. The resulting hiatus was reflected not only by a drop in the number of my publications but also in their significance. Because I was then not far from 60, many people thought I had been overtaken by Father Time, and I ran into problems getting my research support continued. Fortunately, the Air Force Office of Scientific Research (AFOSR) stuck by me.

We should of course have been able to call on the UT Computation Center for help. However, the center was one of the institutions that went downhill with Norman's departure. We had already had an almost unbelievable demonstration of its decay. The Computation Center had to write an operating system for the CDC6600 when UT first got it. However, when a far better one became available from CDC, UT's Dark Ages had set in and the Computation Center was anxious to keep all its staff. They therefore revised their own system instead. Having tested the new version for several weeks, at night and on weekends, they decided it was fine and installed it. Indeed, they not only installed it but also deleted the tapes of the old one. The new system naturally proved to be full of bugs. Nobody, including ourselves, was able to run any calculations for 6 months. Nobody outside UT really believes this story. Unfortunately for me, it was true. Furthermore, nobody was fired for perpetrating this monstrous stupidity, and the Computation Center continued to write, or try to write, its own operating systems. Indeed, when the CDC6600 finally had to be replaced, the Computation Center succeeded in getting it replaced by another CDC computer, one that was already obsolete but had a similar architecture, so that they could continue to write the operating system for it. Even the staff at the Computation Center doubted their ability to write an operating system for any other computer!

After these two disastrous administrations, it did look for a time as though my perseverance might pay off because the next president, Peter Flawn, turned out to be as good at the job as his two predecessors had been bad. However, he lasted only 4 years, and his successor proved to be another lemon.

During those 4 years, Peter did a wonderful job. He started by replacing all the vice-presidents and deans and was well on the way to cleaning up the whole administration. However, he then resigned as president and reverted to being a professor of geology. Peter did not at the time give his reasons for resigning as president. I think he did so at the point when something had to be done about the other and more serious problem, namely interference by the state legislature and the state board of control. When Peter accepted the presidency, he announced publicly that he was doing so only on the understanding that his assignment was to raise UT to the front rank, and that if he felt at any time that the regents were not giving him full support in this objective, he would resign. I suspect that this in fact happened when he began to tackle the problem of state interference. The regents at UT are appointed personally by the governor of Texas.

At the 1973 Bürgenstock Conference. Left to right, R. Breslow, Dewar, Sir John Cornforth, and J.-M. Lehn.

Life in Austin

Apart from my problems with UT, we enjoyed living in Austin.
We had a very beautiful house on the edge of a canyon, looking
out over miles and miles of wooded hills. When we first went
to Austin, our house was right at the edge of the city. Indeed,
all our friends thought us crazy to live so far away from everyone else. By the time we left, the city had expanded several
miles beyond us. However, although the hill tops were built
over, the valley behind our house remained undeveloped, so our
view was not spoiled. Indeed, the net effect reminded visitors
of hilltop villages in France or Italy. Texas appeals to the British.
UT swarmed with them. Friends of ours once held an open
party on the night of the Oxford—Cambridge boat race, inviting
anyone to come who had been to either university. They were
a bit surprised when 150 people turned up! Now that air condi-
tioning is universal, the Texas climate seems like heaven to anyone reared in Britain, and the basic niceness, open friendliness,
and good manners of Texans come as a joy to anyone who has
suffered from the rudeness and bad manners one encounters so
often in Britain. Similar comments apply, though less strongly,
to comparisons of the South with New England or the Midwest.
The South is a very pleasant place to live.

When we first went to Austin, it was admittedly backward in many ways. The shops were poor and the food shops
and restaurants a disaster. Mary had to do all her shopping for
clothes in New York, and we even had to import cheeses and
oil. However, as time went on, things improved rapidly, and a
few years ago Austin had become an almost ideal place to live.
At present, it is suffering badly from the recession, like Texas
generally, and it is also now beginning to get too big, the popu-
lation of metropolitan Austin having increased threefold since
we went there in 1963.

Mary was barred from a teaching appointment at UT by
the nepotism rules. However, after we moved to Austin, UT
appointed her a research associate, with full faculty privileges.
This appointment was a very pleasant change. She was also fortunate in that Austin had one of the leading American historians
in her field, Stan Lehmberg, who moreover welcomed her
arrival. English Tudor history is naturally not a popular field in

Robert and Steuart in Austin in 1961.

America because almost all the sources are in England. Long visits to England are therefore essential for anyone carrying out research. A further benefit was that the library at UT was not only one of the 10 best in America but also, thanks to Stan's efforts, strong in the books Mary needed for her work.

Mary's need for visits to England was also met some years later when we bought a flat in London to make visits to Europe easier. I also needed to go to Europe frequently to give lectures and attend conferences. At that time, there was still a 40-lb weight limit for baggage on flights across the Atlantic, including hand baggage, with massive charges for any excess, so it was a great help to us not to have to carry everything to and fro. We also needed to visit England because we both had parents there who lived to be very old. As it turned out, we bought the flat at just the right time. Usually, in situations like this, we are either too early or too late. Property values in London, which had not changed much since we left, suddenly started to rise. Six months after we bought the flat, they had doubled, and after a year they had tripled. If we had waited even a year to buy it, we would not have been able to afford it, and it proved more-over an unexpectedly good investment.

Being able to visit Europe frequently was also a blessing in other ways. The Dark Ages had just descended on UT, and being able to get away from Austin made the situation much easier. Also, one of our major interests has always been medieval architecture and medieval sculpture, in particular misericords, the carvings underneath choir seats in medieval churches and cathedrals, and we were able to see all the major examples in our travels around Europe. Some day we hope to go through our huge collection of photographs in detail and perhaps write a book.

A few years earlier, in response to the growing problem of hiring faculty with academic wives, UT had abolished its nepo-tism rules. Mary was asked if she would like to have a normal faculty appointment. By this time, however, we were at last free from all family commitments in Austin, both Robert and Steuart being married, and Mary had been coming with me on all my trips. She said flatly that she was not about to stay in Austin teaching courses in European civilization while I was in London, Paris, or Rome. So she remained a research associate. I was relieved by her decision because I would not have liked to start traveling alone again. She certainly chose correctly. Soon after-ward, when Norman Hackerman left, the history department at UT fell apart, and it has not recovered.

Research at Austin

For the first few years, a large part of my research was con-cerned with borazaromatic compounds. Although this work led to much new information concerning their chemistry, the syn-thesis of a large number of new aromatic ring systems, and pub-lication of a large number of papers, it did not lead to any dramatic new developments. As I have said already, bora-zaromatic compounds are really too normal to be exciting. This is perhaps the reason for their otherwise-surprising neglect. Usually, the discovery of a new aromatic ring system leads to a flurry of activity. Here, a whole new *class* of aromatic com-pounds has been virtually ignored!

One result of considerable theoretical interest did, however, come from studies[56] of electrophilic substitution in derivatives of our first borazaromatic compound, 10,9-borazarophenanthrene (**39**). The most reactive position was found to be C-8, followed closely by C-6, and then C-2, exactly as predicted by PMO theory.[56] Frontier orbital (FO) theory, on the other hand, predicts C-8 to be the one-but-least reactive position in the whole molecule! Furthermore, the position with the highest frontier orbital coefficient is C-3, which, in fact, seems almost completely inert. Although this is only one of many cases where FO theory fails, it is certainly one of the most spectacular. I will return to this theme later.

Another interesting result of our research was our discovery that the coordination number of boron can be determined unambiguously by ^{11}B NMR spectroscopy.[57] We were thus able to show that hydroxyboron derivatives, including boric acid itself, act as Lewis acids, not protic acids, by adding hydroxide instead of losing a proton. Indeed, the only BOH compounds that we found to act as protic acids were the corresponding derivatives of our borazaromatics, addition to boron being inhibited by the corresponding loss of aromatic stabilization.

Perhaps our most significant result, which, however, has been largely ignored, came from a further study of the complexes formed by aromatic compounds with "acceptors". We had previously found[51] evidence that Mulliken's description of these complexes as charge-transfer complexes was incorrect. We confirmed this hypothesis by comparing[58] the frequencies (ν) of the charge-transfer bands in the spectra of a number of such

39

complexes with the equilibrium constants (K) for their formation. The complexes were derived from a variety of aromatic compounds with a common donor. If Mulliken's theory was correct, there should have been a linear relation between ν and log K. Our results showed no semblance of such a relation. This result had escaped notice simply because equilibrium constants are much harder to measure than spectra. The only previous measurements had been for compounds so closely related (e.g., methyl derivatives of benzene) that only the charge-transfer contributions varied. We used a wide variety of molecules as "donors".

Mulliken's ideas gained acceptance largely because of the belief that complexes of this kind are associated only with observable charge-transfer transitions, and the fact that such transitions are observed only for complexes between "donors" and "acceptors". Because charge-transfer transitions are weak, they can be observed only if they lie at lower energies than the locally excited transitions of the components. Furthermore, the separation of charge during such a transition tends to increase its energy. Charge-transfer bands can therefore appear at longer wavelengths only if the energy of the donor MO is high and that of the acceptor MO low. This situation is possible only if one component in the complex has a low ionization energy (i.e., is a good donor) and the other has a high electron affinity (i.e., is a good acceptor). The fact that most of the complexes so far reported have been of donor–acceptor type is therefore simply an artifact of the method used to discover them.

Indeed, another of our contributions, this time purely serendipitous, was a new procedure[59] for detecting such complexes. This came out of a mechanistic study[60] of the solvolysis of β-arylethyl esters. In the course of this study, we prepared 1-pyrenylethyl tosylate (41) and measured its proton NMR spectrum. To our surprise, the methyl signal was displaced upfield by 1 ppm. The displacement was, moreover, temperature-sensitive, decreasing at higher temperatures. The analogous aryl-methyl ester was normal. The only reasonable explanation was that 41 folds up into a "sandwich," the tosylate and pyrene moieties forming an intramolecular complex in which the methyl group is shielded by the diamagnetic pyrene ring. Indeed, similar shifts were observed in a solution of a mixture of pyrene and methyl tosylate.

40   **41**

The project[60] that led to this discovery was also of interest. β-Arylmethyl halides and esters can, in principle, solvolyze either by conventional S_N2 or S_N1 mechanisms, or by an alternative S_N1 mechanism involving π complexes (**40**) as intermediates. Our object was to distinguish between these mechanisms. We did this by the approach used earlier[37] for normal S_N2 and S_N1 reactions, that is, by studying the rates of solvolysis of species that contain different aryl groups derived from even AHs ($\pm E$ substituents),[28,48] (e.g., vinyl, phenyl, naphthyl). Because such substituents are nonpolar, they influence reactions only by mesomeric (resonance) interactions. Because the delocalized systems in the transition states (TSs) corresponding to different mechanisms differ, so also do the effects of $\pm E$ substituents, and because the interactions involve AHs, they can be estimated simply and effectively by PMO theory.

We were thus able to demonstrate that the reactions can take place either by the S_N2 mechanism or by the mechanism involving π-complex intermediates. The rates of the S_N2 reactions should have been, and were, independent of the substituent, whereas the rate of a π complex reaction should be greater, the greater its $\pm E$ activity.[28,48] As expected, a logarithmic plot of the rate against $\pm E$ activity consisted of two straight lines, one horizontal, corresponding to species reacting by the S_N2 route.

This approach, which we termed the *alternant hydrocarbon* (AH) method, has several advantages over the conventional counterpart in which substituted phenyl substituents are used and the results are interpreted by using the Hammett equation. In particular, the fact that polar interactions are not involved greatly increases the discrimination between different mechanisms.

I will briefly mention a few other projects that followed on from earlier work.

A return to the liquid-crystal arena led to the discovery that nematic liquid crystals can be used very effectively as stationary phases in gas chromatography to separate molecules by shape.[61] This is now a standard technique. Even *m*- and *p*-xylene can be separated without difficulty in this way. The need for liquid crystals stable enough for the purpose led to the first systematic exploration of the ester group (–COO–) as a linking group.[62]

A return to NQR spectroscopy, using another home-made spectrometer, led to the first systematic study of organoaluminum compounds by this technique[63] and some interesting conclusions concerning the electronic structures of bridged dialuminum compounds.

We also built two UV photoelectron (UPE) spectrometers. The first, a low-resolution grid-type instrument, was used in one of the early surveys of organic compounds.[64] The later high-resolution spectrometer was used initially in studies of compounds of elements from higher periods, the most interesting result being a demonstration of the large effect of $p\pi-d\pi$ interactions in Group IVB elements.[65] Later, we developed a high-temperature inlet system so that we could study the UPE spectra of transient species produced by thermolysis.

Our studies of large organic radicals were the first in which the higher ionizations were located and assigned.[66] We were also the first to report[67] the UPE spectrum of benzyne. The latter was a minor coup in that a number of leading UPE spectroscopists had tried and failed. Our results explained why. The UPE spectrum of benzyne is so similar to that of benzene that it took us some time to establish that we had, in fact, observed it! We were of course assisted in all this work by the ability of MNDO (*see* below) to reproduce ionization energies.

A Practical Interest in Theory

I shall now turn to our theoretical work. From the beginning, my interest in theory was purely practical. My own experimental studies of reaction mechanisms had made me increasingly sceptical concerning them. One cannot observe how a reaction takes place because reactions take place too rapidly. The time involved is too short for the intermediate phases to be observed. Our inability to observe reactions is, moreover, due not to the limitations of our current experimental techniques but to the restrictions set by the uncertainty principle. The reaction mechanisms stated so confidently in organic texts have, in fact, a very shaky basis.

I had come to doubt whether the mechanism of *any* organic reaction was really known. Quantum theory provided the only hope of a solution, and in 1960 this seemed far removed from reality. Current "ab initio" methods were limited to very inaccurate calculations for very small molecules. It seemed unlikely that this approach could lead to useful results for reactions in the foreseeable future, certainly not in my lifetime.

Subsequent events have proved me right. Although advances in computers now allow much more sophisticated ab initio calculations to be carried out, and although hundreds of millions of dollars have been spent on such calculations, there is still no prospect of reaction mechanisms being predicted a priori by quantum mechanical calculation. The errors in the energies calculated for molecules are still enormous by chemical standards, amounting to hundreds of kilocalories per mole even for molecules as small as benzene. Chemical behavior admittedly depends on relative energies (e.g., heats of reaction or activation) rather than absolute energies, so a procedure that fails to reproduce absolute energies may nevertheless reproduce relative ones through a cancellation of errors. Even though there is no rigorous reason why the errors should cancel, current procedures might still be useful in an empirical sense, in areas where tests had shown the errors to cancel. Tests of this kind have, however, shown that most ab initio methods fail to give reliable results for reactions. The only exceptions are state-of-the-art ones that have never been adequately tested because of the cost. Even if some procedure of this kind proved adequate, it would be of little practical use to chemists at present because it could not be used to study reactions of large molecules. The situation could be changed only by a huge increase in the speed of computers, larger than anything likely to be attained before the end of the century, or by the development of some fundamentally better ab initio approach.

Isn't chemistry fun? This could be anywhere, about 1970.

However, as I have already mentioned, early studies in
Chicago had convinced me that the problem could be solved by
a parametric ("semiempirical") approach based on an all-
valence-electron SCF (self-consistent field) MO treatment. The
availability of free computer time made it possible for us to start
work on this project. At first, the project formed only a small
part of my research effort; it was also hampered by the fact that

it did not appeal to graduate students. The theoreticians among them had had a conventional quantum mechanical training that made them uninterested in the kind of things we were doing while applications to organic chemistry were, at first, very limited.

For many years, the work was therefore carried out by a succession of (fortunately) very able organic postdoctoral students who wanted to learn about MO theory. None of them knew anything at all about quantum theory, let alone MO theory, when they arrived, and none of them had ever had any dealings with a computer. However, in a few weeks they had learned enough in both areas, not only to carry out calculations, but also to write the necessary computer programs. We did, admittedly, run into difficulty at times, early on, through our lack of mathematical expertise. It was some time before I began to get people in my group who had any background in theory.

Our first efforts were necessarily limited to the π-SCF approach we had tried in Chicago. Even these calculations were enough to tax the computers then available, in our case a CDC 1604. Although the CDC 1604 was a good computer in its time, a little slower than the "state-of-the-art" IBM 7090, it could not have been used for all-valence-electron calculations for any but the simplest molecules.

My initial enthusiasm for the π-SCF approximation was, however, justified. In its final form,[68a] it gave heats of formation for conjugated and aromatic hydrocarbons that agreed with experiment to within the limits of error of the best modern thermochemical measurements. The only exceptions were cases involving angle strain, which was not allowed for in our approach, and compounds with large rings, which later work has shown to be outside the scope of the Hartree–Fock (HF) approximation. Later, we extended this treatment to compounds containing heteroatoms, and we also developed an alternative version.[68b] Although this approach was overshadowed by the development of all-valence-electron methods, it may be headed for a revival, now that calculations of this kind can be carried out very easily with a personal computer, even for large polycyclic systems. It should prove useful as a teaching aid, and it would also provide a much better treatment of conjugated systems in molecular mechanics than the ones currently used.

Five years after I came to Austin, the university acquired the first properly functioning example of what was then the world's fastest computer, the CDC 6600. We therefore set out to include all the valence electrons in our calculations, by using parameterized versions of the ZDO approximations [INDO (intermediate neglect of differential overlap) and NDDO (neglect of diatomic differential overlap)] that Pople had introduced.

Our approach was not, in itself, novel. It had indeed been tried by Pople. The treatment he developed [CNDO (complete neglect of differential overlap)] was, however, grossly inaccurate. His failure led to a general belief that the semiempirical approach had no real prospects. Our main achievement has been to show that the fault lay not with the procedure but with lack of effort in applying it. Here we had the advantage of being organic chemists. In organic chemistry, one never expects any reaction to succeed the first time one tries it; however, if one is sure that it ought to work and spends enough time and effort on it, one can usually, in the end, get an 85% yield. This was a lesson I had learned by bitter experience over tropolone! So we simply followed the same approach. I was sure that a semiempirical treatment could be made to work, so we made it work.

Our first two treatments, PNDDO[69] (partial neglect of diatomic differential overlap) and MINDO/1,[70] (modified intermediate neglect of differential overlap, first version) were based on assumed geometries, being parameterized to reproduce energies. This they did remarkably well. Indeed, if the object is to estimate heats of formation of molecules with known geometries, this can be achieved both cheaply and very effectively by using either of these procedures. Calculating geometries as well as energies proved much more difficult. The development of our first really successful treatment (MINDO/3)[71] took many years of effort on the part of exceptionally able colleagues and many millions of dollars' worth of computing time. Since then, we have developed second-generation [MNDO[72] (modified neglect of diatomic overlap)] and third-generation [AM1[73] (Austin model 1)] successors, both with immense effort and at great expense. Current experience suggests that the systematic errors in the earlier treatments have been overcome in

AM1, and that it is at least comparable with ab initio procedures that require many thousands of times more computing time. A general account of this work is given in ref. 74.

We have been alone in trying to develop effective, reliable, and general molecular models based on a parametric approach. Other theoreticians have concentrated on the so-called ab initio SCF approach introduced independently by Roothaan[75] and Hall,[76] an approximation to the Hartree-Fock (HF) SCF MO method in which MOs are approximated by linear combinations of AOs [LCAO (linear combination of atomic orbitals)] approximation.

The term "ab initio" was originally applied to the Roothaan–Hall (RH) approach through an amusing accident. Parr was collaborating in some work of this kind with a group in England. In reporting one of his calculations, he described it as "ab initio", implying that the whole of that particular project had been carried out from the beginning in his laboratory. The term, unfortunately, became generally adopted for all calculations of this kind, unfortunately because it carries a quite unjustified subliminal impression of reliability and accuracy that has been ruthlessly exploited.

Even the HF method itself, let alone the RH approximation to it, is hopelessly inaccurate in a chemical sense. The error in the energy calculated for an organic molecule is of the same order as its heat of atomization, over 1000 kcal/mol in the case of benzene. The errors can be reduced by as much as three-quarters in state-of-the-art methods in which partial allowance is made for electron correlation; however, the errors are still enormous by chemical standards. Although these procedures may provide chemically useful predictions, they can do so only through a cancellation of errors, and the existence of such a cancellation can be established only by experiment.

Current "ab initio" procedures can therefore be relied on only so far as they have been tested by comparison with experiment. The fact that they are "ab initio" provides them with no special dispensation. The real distinction is not between ab initio and semiempirical but between a priori and empirical, an a priori procedure being accurate enough to lead to unambiguous predictions without reference to experiment. This, indeed, is the distinction that was drawn in the early days of quantum theory. Calculations that are accurate a priori in a chemical sense are currently limited to molecules with, at the most, half a dozen electrons.

As long as our work was limited to π systems, or to preliminary treatments with obvious defects, those in the ab initio area regarded it with amused tolerance. However, when we finally began to rival ab initio procedures with ones requiring a thousand times less cost in computing, all hell broke loose. I had never previously had trouble with referees. Indeed, I had always welcomed their comments and criticisms. Now, however, comments and criticisms were replaced by torrents of childish and often personal abuse. Every effort was made to suppress publication of my papers and papers by others reporting calculations that used our procedures, using every dirty trick in the book. Even now, when organic chemists all over the world are using our methods, when they are in use in virtually all industrial laboratories, and when the better "ab initioists" are also using them as an aid in their ab initio calculations, we still have trouble from a hard core of recalcitrants.

Because even the best ab initio procedures are still hopelessly inaccurate in an a priori sense, reaction mechanisms cannot be predicted by calculations alone. Equally, they cannot be determined by experiment because reactions take place too rapidly for their course to be observed. Moreover, this restriction is set not by our limited experimental capabilities but by the uncertainty principle. However, even an approximate theoretical treatment can usually limit the possible mechanisms for a reaction to a small number of alternatives, and experiments can then usually be devised to distinguish between these alternative possible mechanisms. Indeed, the necessary experiments have often already been carried out. This is the approach we have followed.

In order to apply this approach, one must be able to carry out calculations for molecules for which experimental data are available. These molecules are commonly rather large. Although the best ab initio methods are clearly better than ours, they cannot be used in this way because they are restricted to reactions of very small molecules. While ab initio calculations for more complex molecules are being reported in the literature, they have been made possible only by use of inadequate procedures or unacceptable simplifying assumptions. There is clearly little point in using a procedure that requires thousands of times more computing time than ours do if it is no better than ours, let alone one that is inferior. One of our problems has been the publication of very bad ab initio calculations that are claimed to have "refuted" our conclusions on the grounds that ab initio methods are ipso facto better. Any advertising agent would be proud of inventing the term "ab initio".

Our parametric approach is not limited to the approximations (INDO and MNDO) that we have used hitherto. Indeed, because faster computers are now available, we have already started work on treatments based on better approximations. Such an approach, based on a procedure including all overlap and using a large basis set, could well provide really accurate results.

The reliability of our conclusions has naturally increased as we developed progressively better approximations. It has also become progressively easier to predict their performance in specific connections with the accumulation of information derived from previous work. Any procedure is bound to work better in some situations than in others, and all procedures have their areas of weakness. A major advantage of the procedures we have developed is that each of them has been used in so many different connections, and applied to such a wide variety of molecules, that it is usually possible to assess in advance its likely performance in some new application. This is the one good result that came from our past treatment as pariahs by the ab initio community. Their hostility frightened others from trying to follow our lead or "improve" on our parameters. As a result, I have been able to keep the number of procedures to a minimum, each of which has consequently been used very extensively. We have not released, or even used, a new approximation unless it represented a major advance over existing ones, and we have usually waited until we were sure that it was as good as we could make it. Thus our first two procedures, PNDDO and MINDO/1, represented preliminary excursions where we assumed geometries and calculated only energies. Both in fact were remarkably successful. An approach along these lines would probably give much better estimates of the energies of stable molecules than any other current quantum mechanical procedure.

Our next contribution, MINDO/2, was the first in which we calculated both energies and geometries. Because this development allowed us to study reactions for the first time, it was released before it was fully optimized and hence proved erratic. The fully optimized version (MINDO/3) appeared 3 years later. Our next treatment, MNDO, was based on the NDDO approximation instead of INDO. Although MNDO has proved remarkably effective, it suffers from some serious deficiencies that we were unable to correct by further parameterization. After 8 years, we finally resorted to the brute force expedient of introducing an additional parametric function. I think it unlikely that the result (AM1) could be significantly improved by further reparameterization of the AM1 algorithm. Our earlier procedures are now effectively obsolete except for a few clearly defined situations where AM1 stumbles and MNDO or MINDO/3 has been shown to perform better. Apart from these special cases, I am now beginning to feel real confidence in the mechanistic predictions that AM1 gives.

My attempts to keep the semiempirical approach under control have met with only one failure, a reparameterization of AM1 termed PM3. Whether this will prove generally better than AM1 seems doubtful, and the improvement is certainly unlikely to compensate for the confusion caused by the existence of two essentially equivalent procedures. Similar comments apply to the existence of two similar computer programs for carrying out semiempirical calculations, MOPAC and AMPAC. I would prefer not to comment on the circumstances behind these two unfortunate developments. However, the situation should soon be resolved by a basically new treatment that we are currently developing and that seems likely to prove a clear advance over both its predecessors.

In the ab initio area, on the other hand, chaos reigns. Literally hundreds of different basis sets have been used in ab initio calculations, each defining a different empirical procedure, and similar comments apply to "post-Hartree–Fock" procedures in which allowance is made for electron correlation. Furthermore, no systematic tests or comparisons have been reported, apart from ones by us,[77a] where we compared results from our procedures with those from ab initio calculations using three of Pople's more popular basis sets. Those in the ab initio area still follow precedent by using procedures that have not been tested and by failing to test any new procedures they develop.

The fact that ab initio methods can reproduce only relative energies admittedly makes it more difficult to test them. However, we have not only pointed out[77a] a simple solution of this problem but have also developed a computer program to implement it.[77b] Because this program is written for IBM-compatible PCs and is freely available, there is no longer any excuse for failure to test ab initio procedures. Nevertheless, we alone have published such tests. My comments on the performance of our procedures, relative to ab initio ones, were based on our results.

Semiempirical Procedures

One of the criticisms commonly leveled at semiempirical methods is that they represent no more than methods of interpolation and are useful only in areas and for compounds for which they have been parameterized. This criticism is certainly true of the semiempirical methods that others have developed. The striking thing about ours is that they do *not* merely reproduce the properties for which they were parameterized, nor are they confined to molecules of the kind used in the parameterization. They reproduce *all* ground-state properties of molecules of *all* kinds, including ions, radicals, and carbenes, even though the parameterizations were based entirely on properties of normal neutral molecules. The properties used in the parameterization are primarily energies and geometries but include a few dipole moments and ionization energies. The other properties reproduced include molecular vibration frequencies,[78] entropies, and specific heats,[79] kinetic isotope effects,[80] polarizabilities[81,82] and hyperpolarizabilities,[82,83] ESCA (electron spectroscopy for chemical analysis) chemical shifts,[84] nuclear quadrupole coupling constants,[85] properties of polymers (including their electronic band structure),[86] and electron affinities.[87] Thus our procedures provide a very good representation of the way molecules behave (i.e, they provide good molecular models).

Over the years, our work has led to a number of predictions that have been subsequently confirmed by experiment. Our calculations led, for example, to the prediction[88] that *m*-benzyne (**42**) should be a stable species, being really bicyclo[3.1.0]hexatriene (**43**), and that *p*-benzyne (**44**) should likewise exist as a stable bicyclic isomer, **45**. Both these predictions were subsequently confirmed by experiment.[89,90] Another surprising prediction,[91] now generally accepted, was that "cycloheptatrienacarbene" (**46**) is in fact cycloheptatetraene (**47**). Many other similar examples could be cited.

This success is no accident and it has not been obtained easily. Most people seem to assume that developing an effective parametric procedure involves only a simple optimization of the parameters. Unfortunately, this is by no means the case!

1. The parameters in our treatment appear in parametric functions of unknown form. One has first to find the best forms for those functions.

2. The values of the parameters are chosen to give the best overall fit to experiment for various properties of an arbitrarily chosen set (basis set) of molecules. The results naturally depend on the choice of molecules in the basis set.

42

43

44

45

46

47

3. The set of values found in this way is not unique. The
    optimization starts with a set of assumed values for the
    parameters. The final values obtained depend on the
    values assumed initially. Furthermore, there is no way to
    tell whether the set of values obtained is the best, and
    there is no systematic way to find other sets of solutions to
    the optimization equations. All one can do is try various
    sets of initial values for the parameters in the hope that
    one of them will lead to the optimum result.

4. Even the criterion for optimization is indefinite. Every
    procedure performs less well in some cases than in others.
    How serious each such error is depends on the chemical
    importance of the molecule in question. Thus an error of
    15 kcal/mol would be quite unacceptable in the case of
    benzene but relatively unimportant in cubane. The accept-
    ability of the results given by a given set of parameters is
    therefore ultimately a matter of chemical judgement. It
    cannot be assessed automatically by a computer program.

Because there is no a priori way to choose the parametric
functions or a suitable basis set of molecules and properties, they
can be found only by trial. Each set of parametric functions is
tested by a series of optimizations, with various basis sets and
starting values for the parameters. As can readily be imagined,
the whole process is extremely laborious, hideously frustrating,
and enough to tax the patience of a saint. In other words, it is
much like organic chemistry. In organic chemistry, the first time
one tries a new synthesis, it usually fails. However, if one feels
sure that it ought to work, if one has the knowledge and ability
to make judgments of this type, and if one perseveres hard
enough, one usually achieves success. We just followed the
same approach with parameterization. The two who developed
MINDO/3, Dick Bingham and Donald Lo, came to me with
Ph.D.s in organic chemistry. They tried more than 500 combinations of parametric functions before settling on the ones used in
MINDO/3. It took them 2 years.

I have been almost alone in the theoretical area in making
all our programs freely available, instead of commercializing
them. Personally, I think it improper for academic scientists to

do otherwise, particularly if their work has been supported by grants from federal agencies. However, I would now strongly urge others to maintain control of their programs by copyrighting or patenting them, even if they do not want to make money out of them. I shall certainly follow this course in future.

Calculations for Reactions

My main interest in theoretical calculations has been as an aid in the study of reaction mechanisms. As I have already pointed out, the mechanisms of reactions cannot be determined by direct observation because they take place too rapidly. Such observations would violate the uncertainty principle. Nor can reaction mechanisms be determined a priori by theoretical calculation because current theoretical procedures are far too inaccurate. The best approach is to use a combination of theory and experiment. A set of possible mechanisms is first deduced for the reaction by using a combination of qualitative theory and quantitative calculations. The choice between these is made by comparing experimental values for various reaction parameters (such as heats and entropies of activation and kinetic isotope effects) with values calculated for the various mechanisms.

The cost of adequate ab initio calculations usually restricts their use to the simplest examples of reactions, whereas our semiempirical ones can be applied widely. This is important, for two reasons.

First, the simplest example of a reaction is often atypical, as Ingold pointed out some years ago. Ethyl chloride, for example, is the only alkyl chloride that does not give an olefin on treatment with alcoholic potash, and methylamine is the only primary aliphatic amine that does not give an alcohol on treatment with nitrous acid. These examples were previously given in elementary organic textbooks.

Second, because the errors in calculations for related molecules are likely to be the same, any procedure is likely to give better estimates of the relative values of a given property for a number of similar systems than it does of the absolute value of the property for an individual system. Thus any theoretical procedure should give better estimates of the relative values of such things as activation parameters for a number of related reactions than it does of their absolute values for one individual reaction. Inclusion of such data, in particular substituent effects, also greatly increases the number of possible comparisons between theory and experiment.

For both these reasons, calculations should be carried out for a number of examples of a given reaction for which experimental data are available for comparison. Our procedures can be used freely in this way, even for reactions of quite large molecules, and we ourselves have carried out calculations of this type for a very large number of reactions. The only ab initio studies of this type so far reported have been based on inadequate ab initio procedures.

Most of our more significant contributions have naturally come in recent years. We are now harvesting the fruits of the 20 years of unrelenting effort it took to develop our current procedures. Here I will summarize only a few of our more striking results. Much of our earlier work was reviewed in a lecture I gave in 1975, on receiving the Chemical Society's Robinson Medal, which was published[92] in *Chemistry in Britain*.

Nonclassical Carbocations

One major investigation, which has continued over the years, has been concerned with the properties and structures of "nonclassical" carbocations. The large majority of such ions can be represented[15] as π complexes. Problems arise, however, in attempts to study such species theoretically. Both MNDO and AM1 fail to reproduce their properties satisfactorily, and calculations using ab initio methods are satisfactory only if carried out at a very high level. Curiously enough, MINDO/3 performs well in this connection, having been shown to give results similar to those from high-level ab initio methods for a number of simple nonclassical carbocations.[98] Indeed, MINDO/3 still represents the method of choice for carbocations that are too large for study by adequate ab initio procedures.

One example[99] was concerned with the structures of the $C_4H_7^+$ ions formed by ionization of cyclopropylmethyl or cyclobutyl derivatives. The evidence indicates that there are two isomeric ions of this type, both unexpectedly stable, which can interconvert readily. These species have been formulated as cyclopropylmethyl cation **48** and cyclobutyl cation **49**, the latter being the lower in energy. However, while MINDO/3 predicted corresponding minima on the potential energy surface, the former "**48**" proved to be a π complex, **50**, stabilized, as indicated, by back-coordination. The "cyclobutyl cation" proved to be hinge-protonated bicyclobutane **51**, which was predicted to be the more stable, in agreement with experiment. We have recently[100] elucidated the role of such species in the enzymatic synthesis of squalene from farnesol, a key step in the biosynthesis of steroids.

Biomimetic Cyclizations

Another reaction of importance in biochemistry (e.g., the biosynthesis of steroids from squalene) is ring formation by intramolecular addition of a carbocation to a C=C bond. Reactions of this kind are commonly termed *biomimetic cyclizations*. A crucial question is the nature of the intermediate formed by the addition. Is it a classical carbenium ion or a π complex? While I was at QMC, Martin Ansell tried to answer this question by studying the stereochemistry of ring formation in such reactions. He made considerable progress, his most amusing result[101] being the conversion of the phenylalkadienol (**52**) to the chrysene derivative **53** in 60% yield upon treatment with acid. The techniques available at that time were, however, inadequate, and no further work was carried out on the problem for many years. Such reactions are now known to take place by *trans* addition, implying that the intermediates are π complexes rather than classical carbenium ions.[13,15,48] MINDO/3 calculations[102] for some typical examples have confirmed the π-complex mechanism.

48   **49**   **50**   **51**

52   53

Calculations of this kind should also be useful in mass spectrometry for interpreting gas-phase reactions of ions. We have studied two such processes of interest to mass spectroscopists, namely 1,2 eliminations of H_2[103] and the conversion of the toluene radical cation to tropylium.[104] The results in both cases agreed with experiment and led to firm predictions concerning their mechanisms.

The Mechanisms of Pericyclic Reactions

One of my most extensive projects, now almost complete after 15 years, has been concerned with studies of pericyclic reactions, an area with a curious and instructive history.

The term "pericyclic" was introduced by Woodward and Hoffmann[93] to describe a type of reaction in which a series of bonds switch cyclically around a ring of atoms. Although mechanisms of this type had been considered for several reactions, nobody had previously suggested a term for them, and the earlier work had been largely forgotten. As a result, Woodward and Hoffmann's claim that such reactions involved "a new principle of bonding", and their deduction of a set of rules for them, aroused major interest, and their conclusions have been widely accepted.[93] Curiously enough, a simpler and more effective interpretation had in fact been published many years earlier. Because I probably know more about this affair than any other chemist who is still active, I will start with a brief account.

The Diels–Alder reaction was the first pericyclic reaction to attract detailed mechanistic study. In the early 1930s, two mechanisms were under consideration. One of these involved what would now be called a synchronous pericyclic mechanism in which three CC bonds switch cyclically around a ring of six carbon atoms, forming two new CC bonds simultaneously. The other involved a two-step mechanism in which the new bonds form sequentially, formation of the first giving rise to an intermediate biradical. Controversy concerning the choice between them formed quite a large part of a Faraday Discussion in 1936.

MINDO/3 triumphs at a Dewar group seminar. See J. Am. Chem. Soc. *1975, 97, 4439.*

In 1938, M. G. Evans (M. G. as he was known to his friends) produced a typically brilliant and very persuasive solution of this problem,[94] based on an analogy between the MOs in TSs for reactions of this type and the π MOs in conjugated hydrocarbons. In each case, each carbon atom uses three of its AOs to form σ bonds to neighboring atoms, the remaining AOs interacting to form multicenter MOs. Figure 2a illustrates this analogy for the example considered by M. G., namely the Diels–Alder reaction between ethylene and butadiene. In the synchronous TS, the six relevant AOs overlap cyclically, as do the six carbon $2p$ AOs in benzene, whereas the nonsynchronous TS exhibits a corresponding relationship to 1,3,5-hexatriene. Benzene is more stable than 1,3,5-hexatriene, being aromatic. M. G. concluded that the synchronous TS should be more stable than the nonsynchronous one, being likewise aromatic. Conversely, in the seemingly analogous cyclodimerization of ethylene to cyclobutane, the synchronous TS would be a corresponding analogue of cyclobutadiene and hence antiaromatic (Figure 2b). The reaction should therefore be less facile than the Diels–Alder reaction, which it is, and it should take place nonsynchronously. At that time, it had not yet been established that cyclobutadiene is antiaromatic, that is, destabilized by cyclic conjugation. It was, however, generally recognized not to be stabilized. The fact that cyclobutadiene is even less stable than M. G. supposed clearly does not affect the validity of his argument concerning the synchronicity of the Diels–Alder reaction.

M. G. was excited by this discovery. He was, however, prevented from following it up, as he had intended, first by World War II and then by his early death. As a result, his work was overlooked by many chemists in the postwar era, including Bob Woodward and Roald Hoffmann.

When the first paper[95] by Bob and Roald on the mechanism of ring opening in cyclobutene appeared in 1965, I first heard about it at my research seminar in Austin because another member of my group got his copy of *JACS* before I did. In this article, Bob and Roald interpreted the reaction in terms of frontier orbital theory. I immediately pointed out that the preference for conrotatory ring opening could be explained very simply in terms of M. G.'s theory of aromatic and antiaromatic TSs.

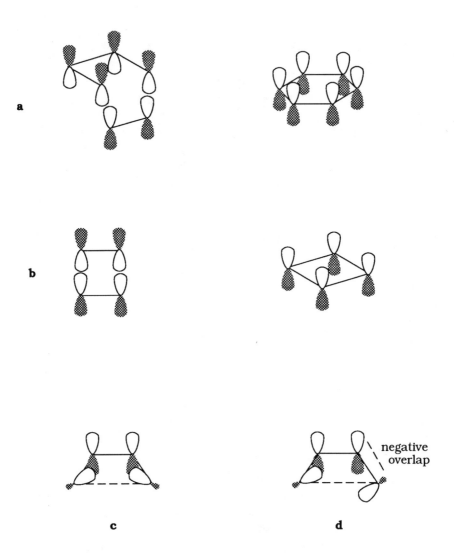

Figure 2. Demonstration of the parallel (a) between the Diels–Alder synchronous transition state and benzene; and (b) between the transition state for synchronous dimerization of ethylene and cyclobutadiene. Transition states for ring opening in cyclobutene: (c) disrotatory and (d) conrotatory.

The reaction can take place in two geometrically distinct ways via TSs that each contain a cyclic array of four carbon AOs and hence a set of cyclic four-center MOs. It might appear, at first sight, that both TSs should be antiaromatic, like cyclobutadiene. However, as Craig[96a] and Heilbronner[96b] had independently shown, there are two different types of aromatic system, Hückel and anti-Hückel, the rules for aromaticity in anti-Hückel systems being diametrically opposite those in the normal Hückel ones. The distinction between the two types depends on whether the phases of the AOs can be chosen so that the overlap integrals between pairs of adjacent AOs are all positive, as is always the case in normal conjugated hydrocarbons. In a Hückel system, they can all be positive; in an anti-Hückel system, at least one integral is negative. As Figure 2c shows, the TS for disrotatory opening in cyclobutene is of Hückel type and hence antiaromatic, whereas the conrotatory TS (Figure 2d) is of anti-Hückel type and hence aromatic. Conrotatory ring opening is therefore preferred.

Producing this explanation required no originality on my part. It should have been obvious to anybody who knew M. G.'s work and had read the papers by David Craig and Edgar Heilbronner. David's examples of anti-Hückel systems involved phosphonitrile chlorides and might have been over-looked by organic chemists, but Edgar's ingenious suggestion concerning Moebius cyclic polymethines had appeared in *Tetrahedron* and had had wide publicity. I was familiar with both.

Why did I not point all this out immediately in print? Because it was not something for which I could claim any credit and because it never occurred to me that someone else would not soon draw attention to M. G.'s work. Also, I had a large backlog of papers to write, reporting my own contributions. However, given that I had been a close friend of M. G.'s and he was dead, I should have intervened. I was finally shamed into doing so when Howard Zimmerman "rediscovered" M. G.'s interpretation. By this time, however, "orbital symmetry" was deeply entrenched in the literature.

This episode raises an important point. In any new development, the crucial step is the introduction of some entirely new principle or idea. Discovering the new principle is the difficult part. Once a new principle or idea has been published in enough detail to be intelligible, no real ability or originality is needed to work out its consequences. Because new ideas are rare, most people who have one are only too happy to follow it up, usually in all possible detail. Indeed, in many cases the discoverer of a significant new idea turns its study into his or her life's work. As a result, new ideas tend to be ignored if, for any reason, their inventors do not follow this course. If someone else then publishes a series of papers describing applications of the idea to specific problems, he or she is likely to be given credit not only for the applications but also for the idea itself. In such circumstances, few refuse the unjustified credit! Indeed, human nature being what it is, the person in question often comes to believe that he or she really was responsible for the idea. I have suffered myself more than once in this way; perhaps the π complex theory of metal–olefin complexes is the most striking example.

M. G.'s death was a major loss to chemistry as well as to his many friends. He was not only one of the most able and original chemists of his generation but also one of the kindest. He went out of his way to help others, including many who, like myself, had no claim on him. He had a wonderfully positive attitude to life combined with total integrity and a complete lack of conceit. British chemistry now would be very different if he had lived longer. He was a victim of smoking at a time when its dangers were not yet known.

Returning to pericyclic reactions: When our first really effective procedure (MINDO/3) became available, one of our first uses of it[97] was to study the relationship between the interpretation of pericyclic reactions in terms of orbital symmetry, and the interpretation in terms of the aromaticity of transition states (TSs).

Our first concern was to check the role of orbital symmetry. If this were the determining factor, the Woodward–Hoffmann rules should apply rigorously only in cases where symmetry is preserved during a reaction. Yet as Bob and Roald had themselves shown, the rules seem to apply regardless of whether the reactants in a reaction have symmetry. We quickly found that "forbidden" reactions do in fact involve crossings of occupied and unoccupied MOs, regardless of whether the systems have symmetry, regardless of whether symmetry is retained during the reaction, and, indeed, regardless of the route followed by the reaction. The MOs in a molecule must therefore be distinguished by some factor other than symmetry that can nevertheless prevent them from mixing. We supposed that the determining factor must be related in some way to the topology of the nodes in the relevant MOs, an idea that is now widely accepted. However, the nature of the relationship is still not clear. In accordance with this quasi-topological distinction, the MOs can be divided into sets, such that two MOs from different sets cross freely whereas two MOs from the same set exhibit an avoided crossing. A collection of isomeric molecules can therefore be likewise divided into groups, such that the number of occupied MOs in each set of MOs is the same for two molecules in the same group. Members of the same group are termed *homomers,* and members of different groups are termed *lumomers.* Homomers can thus be interconverted without crossing of an empty MO with a filled one and are consequently "allowed" in terms of the Woodward–Hoffmann rules, whereas interconversion of two lumomers involves a HOMO–LUMO crossing and is consequently "forbidden". This distinction applies regardless of the route followed by the reaction, and it can also be shown that the TSs for "allowed" and "forbidden" reactions are, respectively, aromatic and antiaromatic.

The situation regarding symmetry is now clear. The distinction between "allowed" and "forbidden" reactions has nothing to do with symmetry. If the reactants have some element of symmetry and if there is a path for the reaction in which this symmetry is retained throughout, a head count of electrons may allow one to tell that there must have been an orbital crossing and that the reaction must therefore be "forbidden". In most cases, however, the symmetry criterion cannot be invoked. On the other hand, it is always easy to tell whether a given synchronous TS is aromatic because the rules for aromaticity are unambiguous and hold generally.

According to the Woodward–Hoffmann rules, "forbidden" reactions never take place in a synchronous manner, whereas "allowed" reactions are always synchronous. This conclusion rested on the assumption that the distinguishing factor is symmetry and that symmetry controls pericyclic reactions as rigorously as it does transitions in spectroscopy. According to M. G.'s interpretation, the determining factor is the aromatic energy of the TS that can have any value from large and positive through zero to large and negative. If the aromatic energy is small, other contributing factors may outweigh it, leading to exceptions to the rules. We have indeed found evidence for many such exceptions.

The first was indicated by calculations[105] for the electrocyclic ring opening of bicyclo[2.1.0]pentene (**54**) to cyclopentadiene (**55**). Our results indicated that the reaction is not merely concerted, but synchronous, even though it is "forbidden" and has a relatively low activation energy (26 kcal/mol), which our calculations reproduced almost exactly. Similar results were obtained[106] for the analogous conversion of Dewar benzene (**56**) to benzene, which, however, was predicted to be concerted but not synchronous, the TS being unsymmetrical. The difference between **54** and **56** is not surprising in view of the the lower steric requirements in **56**. These results are consistent with Evans' principle.[94,107]

"Forbidden" pericyclic reactions tend not to be concerted, let alone synchronous, because of the antiaromaticity of the corresponding TSs. This does not, however, mean that such a reaction cannot be synchronous. Other factors that favor a symmetrical transition state may outweigh the effects of antiaromaticity. There is no reason why a "forbidden" pericyclic reaction should not take place very readily by a synchronous mechanism if this is favored sufficiently by other factors. We have encountered a remarkable example. The "forbidden" synchronous cyclodimerization of silene (**57**) to 1,3-disilacyclobutane (**58**) is predicted by MNDO[108] to take place with *zero* activation energy, even by a path in which symmetry is fully retained.

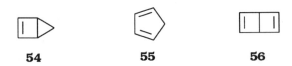

54   **55**   **56**

$$H_2Si = CH_2$$

57

$$H_2Si - CH_2$$
$$\quad | \qquad |$$
$$H_2C - SiH_2$$

58

The assumption, that "allowed" pericyclic reactions are necessarily synchronous in the absence of special factors, also seems to have been refuted by our work. The aromaticity of a concerted TS for an "allowed" reaction certainly tends to make the TS symmetrical, because aromaticity is greatest when the interactions between the participating AOs are equal. There is, however, a second factor to be considered. Although I have drawn attention to this in lectures for many years, I only recently[109] discussed it in print.

Consider a *one-bond* reaction, that is, a reaction where only one bond is broken and one formed. The Evans–Polanyi representation of such a reaction is indicated in Figure 3a. "A" is the bond-breaking curve, representing a plot of energy vs. a reaction coordinate for a process in which the breaking bond is broken. "B" is the bond-making curve, corresponding to formation of the new bond. The crossing point (C) represents the transition state.

Now, consider an analogous *two-bond* reaction, one where two bonds are broken and two are formed. If the bonds break synchronously, the corresponding bond-breaking curve (A' in Figure 3a) will rise twice as high as for the one-bond reaction, and the crossing point (C') will be correspondingly higher. This simple argument shows unambiguously that, other things being equal, the activation energies of synchronous two-bond reactions will be double those of analogous one-bond reactions.

A two-bond reaction will therefore usually prefer to take place in one-bond steps, even if these steps involve a high-energy intermediate (Figure 3b). The steps may, of course, coalesce (Figure 3c). The reaction is then a *two-stage* reaction, concerted but not synchronous. One of the bond-forming–breaking processes (A) takes place between the reactant and transition state, while the other (B) takes place between the transition state and the product.

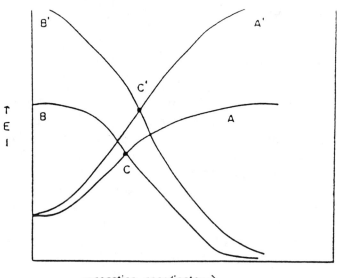

(a)

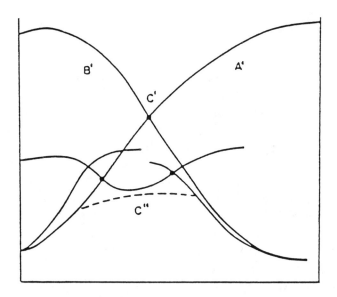

(b)

Two-bond reactions are thus "forbidden" in the same sense that antiaromatic pericyclic reactions are "forbidden". In other words, they tend to have high activation energies. They are synchronous only if some other factor outweighs the unfavorable effect of synchronicity. The same should be true, a fortiori, for reactions in which three or more bonds are formed or broken, a conclusion summarized in what may be termed the multibond rule: *Synchronous multibond reactions are "forbidden".*

In the case of a multibond reaction that is "allowed" by the Woodward–Hoffmann rules, synchronicity is favored by the aromaticity of the transition state. In the formalism of Figure 3, this behavior corresponds to stabilization by resonance between the reactantlike and productlike structures, as indicated by the dotted line in Figure 3b. The resulting decrease in energy may make the corresponding TS (C'') lower in energy than the nonsynchronous TS. Whether this is the case can be determined

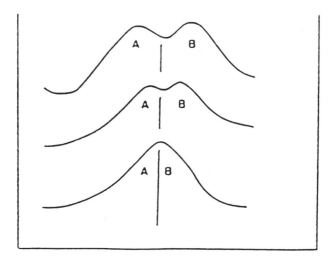

(c)

Figure 3. (a) Evans–Polany plot for analogous one-bond (ABC) and two-bond (A'B'C') reactions. (b) Evans–Polany plot of a two-step reaction and a corresponding two bond reaction, the dotted line indicates the effect of resonance in the transition state of the latter. (c) Merging of the separate steps in a two-step mechanism into a concerted two-stage mechanism.

only by a quantitative assessment of the opposing factors. There is certainly no reason why a pericyclic reaction that is "allowed" by the Woodward–Hoffmann rules should be synchronous if it is "forbidden" by the multibond rule.

The terms "allowed" and "forbidden", applied to reactions, are therefore the unfortunate product of a historical oversight. The terms "aromatic" and "antiaromatic" would have been much better. It is now too late to change, but a reasonable compromise is to enclose the terms "allowed" and "forbidden" in quotation marks, as I have done here, when they are applied to reactions.

Summaries of our work on various specific pericyclic reactions follow.

Diels–Alder Reactions. The Diels–Alder reaction is the archetype of pericyclic processes, and its mechanism has been the subject of continuing interest since it was discovered 60 years ago. In recent years, most organic chemists have regarded it as a typical synchronous pericyclic reaction, involving a symmetrical aromatic transition state analogous to benzene. It has been generally assumed that other analogous cycloadditions follow a similar course.

The first real evidence to the contrary came from an MINDO/3 study[110] of the reaction of ethylene with butadiene, which led to the conclusion that reactions of this kind are *not* synchronous. They take place via very unsymmetrical transition states in which one of the new bonds is almost completely formed while the other is still very weak. Such a species, formed from a biradical by a weak interaction between the radical centers, is termed a *biradicaloid*.

Because the energy of a biradicaloid differs little from that of the corresponding biradical, it should be possible to interpret the course of Diels–Alder reactions by assuming that the transition states *are* biradicals. This indeed is the case, as organic chemists have known for many years. The rates and regioselectivities of Diels–Alder reactions can be predicted with assurance on this basis, a fact that has indeed been quoted in text books as a useful mnemonic for students.

A similar situation applies to other cycloaddition reactions. Indeed, Raymond Firestone[111] suggested some years ago, on this basis, that dipolar addition reactions take place via biradical intermediates. The significance of these results has been ignored in recent years because of the mistaken, but almost universal, belief that "allowed" pericyclic reactions are necessarily synchronous. The situation has moreover been obscured by the assumption that the choice is between a synchronous mechanism and a two-step mechanism involving a biradical as a stable intermediate. The fact that the experimental evidence is inconsistent with the latter possibility has therefore been taken as evidence that such reactions must be synchronous. The evidence is, however, entirely consistent with a third alternative, a mechanism that is concerted but not synchronous and involves a very unsymmetrical biradicaloid transition state.[112] Recent calculations[113] indicate that dipolar additions also take place via very unsymmetrical TSs that are close to biradicals or zwitterions in structure.

The possible role of biradical and biradical-like intermediates has, of course, long been recognized. Indeed, our discussion[95–97] of the relationship between the Evans[94] and Woodward–Hoffmann[93] treatments of pericyclic reactions rested on the biradical-like nature of the intermediates in "forbidden" reactions. The conclusion, that "allowed" reactions can also take place via similar intermediates rather than via aromatic TSs, was, however, novel and aroused much controversy. Our original conclusions concerning the Diels–Alder reactions were admittedly subject to some uncertainty because our calculations had been based on MINDO/3 or MNDO. Our conclusions have been confirmed by a detailed AM1 study[112] of the Diels–Alder reactions of butadiene with ethylene, acrylonitrile, and the dicyanoethylenes, which seems to suggest strongly that none of these reactions, with the possible exception of the first, is synchronous.

Apart from its theoretical interest, this work is of practical significance in view of the growing importance of cycloaddition reactions in synthesis. Their rates, regioselectivities, etc., can be interpreted simply and reliably by assuming that their TSs are close to biradicals or zwitterions in structure. This approach is both simpler and much more reliable than the currently fashionable treatment in terms of frontier orbital theory, a point that I will return to presently.

Our recent work on the Cope rearrangement (*see* below) led us to suggest[112] that Diels–Alder reactions may also take place by alternative aromatic (ARO) and biradical (BR) paths, rather than by a continuous range of mechanisms. Unfortunately, we were unable to establish this by our own calculations because of the inadequate treatment of biradicals by AM1. Recently, however, Bernardi et al.[114] reported an MCSCF ab initio study of the ethylene–butadiene reaction in which they located two distinct TSs, one corresponding to the ARO mechanism and the other to the BR mechanism, the former lower in energy by 2 kcal/mol. Because substituents should favor the BR TS, and because our work on the Cope rearrangement has shown that the MCSCF procedure probably favors aromatic TSs, the calculations by Bernardi et al. suggest strongly that the large majority of Diels–Alder reactions take place by the BR route.

Cope Rearrangement. Another pericyclic reaction where our calculations challenged current ideas is the Cope rearrangement of 1,5-dienes (**59**). Again, it had been almost universally accepted that these reactions were synchronous and took place via symmetrical transition states (**60**) analogous to benzene. Although Doering et al.[115] had suggested that the reactions might take place in steps or stages via 1,4-cyclohexylene biradicals (**61**) as intermediates or TSs, this suggestion had not gained much acceptance.

Our experimental studies[116] of the Cope rearrangements of **59** and several of its phenyl and diphenyl derivatives, together with MINDO/3 calculations[117] for the rearrangement of **59** itself, strongly supported the Doering mechanism. However, the intermediates are not biradicals. Although there is no other way to formulate them using conventional notation, the through-bond coupling between the radical centers is so strong that they behave like normal closed shell molecules. Calcula-

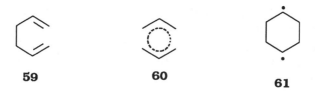

59   **60**

61

tions for them can indeed be carried out by using the normal closed shell version of AM1. The term biradicaloid has been suggested for singlet biradicals in which there is strong coupling between the radical centers. It should be reserved for special cases such as this where the pairing of electrons is comparable with that in normal covalent bonds.

This conclusion was challenged by Osamura and his coworkers on the basis of MCSCF ab initio calculations that predicted the rearrangement of 1,5-hexadiene to be a normal ARO pericyclic reaction. We therefore reinvestigated the matter by carrying out AM1 calculations[118] for the Cope rearrangements of 1,5-hexadiene and a number of its derivatives. Our calculations reproduced the observed activation parameters and kinetic isotope effects with remarkable accuracy, a result suggesting strongly that the predicted BR mechanisms were correct.

One problem, however, remained. Whereas the Cope rearrangement of **59** takes place predominantly via a chair-type TS, as Doering and Roth[119] showed by experiment and as our calculations[117,118] indicated, there is also a slower concurrent reaction involving a boat-type TS. Our calculations gave good estimates of the difference between the free energies of activation for the two paths and of the entropy of activation for the chair reaction, but there was a serious discrepancy between the calculated (−11 eu) and observed (−3.0 ± 3.6 eu) entropies of activation for the boat reaction. Entropies of activation calculated by AM1 are rarely in error by more than 2 eu.

On further investigation,[120] we found that these reactions can each take place in two different ways, via two distinct TSs, one TS of BR-type and the other ARO. The lengths of the forming and breaking bonds were 1.6 Å in the BR TSs and 2.0 Å in the ARO TSs, whereas the entropies of activation for the ARO TSs were half those for the BR ones. Studies of several other Cope rearrangements[121] led to similar results, suggesting that the phenomenon of dual TSs is quite general. The entropy of activation calculated for the ARO boat rearrangement (−6.0 eu) agreed with experiment and with values observed later by Doering and Troise[122] for reactions constrained to take place by the boat path.

These results preceded the calculations for the Diels–Alder reaction noted in the previous section, showing for the first time that a reaction that takes place in one kinetic step may do so via either of two alternative TSs. Here the situation is particularly striking because the geometries of the two TSs are so similar.

We have also carried out extensive ab initio calculations[123] for the BR and ARO Cope rearrangements of 1,5-hexadiene that show that the two TSs cannot both be obtained in the correct energy relationship to one another unless a very large basis set is used (at least double-zeta-plus-polarization), together with effective allowance for electron correlation. Because Osamura et al. used the 4-31G basis set, it is not surprising that they found only the ARO TS.

Implications for Computational Chemistry. These observations have not only established the mechanisms of Cope rearrangements beyond any reasonable doubt but have also raised a new and unexpected obstacle to theoretical studies of reaction mechanisms. Until now, it has been generally assumed that a reaction in which the reactants are converted to the products in one kinetic step can take place in only one way, via a unique TS. This assumption greatly simplifies the problem of calculating mechanisms because if it is true, and if a TS is located, it can be assumed to be the real, unique TS for the reaction in question. The assumption seemed reasonable as long as no exceptions were known. This, however, is no longer the case; exceptions have been found. If the Cope rearrangement can take place via either of two alternative TSs, so too may any other reaction for which more than one TS is possible. Henceforth, in theoretical studies of any reaction, it will be necessary not merely to locate a TS but also to ensure that no alternative TS of lower energy exists. Proving that a TS does not exist requires even more computation than locating one that does. Our results have also shown that ab initio calculations are unreliable in this connection unless they are carried out using a large basis set together with effective allowance for electron correlation. Ab initio calculations that meet all these conditions are at present limited to reactions of small molecules.

Claisen Rearrangement. The Claisen rearrangement of allyl vinyl ethers bears an obvious analogy to the Cope rearrangement and might be expected to follow a similar mechanism. Our calculations[124] for a large number of Claisen rearrangements indicate, however, that this is not really the case. While one or two simple examples show dual mechanisms, the TSs do not show a clear distinction between ARO and BR types, and in the majority of cases there is only one TS of intermediate type.

There is a simple reason for this difference between the Cope and Claisen rearrangements. Whereas the prototype Cope rearrangement, that of 1,5-hexadiene, is thermoneutral, the analogous Claisen rearrangement, of allyl vinyl ether, is exothermic by about 20 kcal/mol. The Claisen TS therefore should be, and is calculated to be, an early one in which the breaking bonds are still short and the forming ones still long. This TS should represent a point on the potential energy surface where the two valleys, corresponding to the ARO and BR reactions paths, have hardly begun to separate.

Benzidine Rearrangement. Another mechanistic problem that has interested me for many years concerns the benzidine rearrangement. I have already described how my early attempts to explain it led to the invention of the π complex theory. Recent improvements in our computer facilities have at last enabled us to carry out meaningful calculations for the conversion of **10** to **11**. Much to my surprise, the reaction turned out to be a simple pericyclic process with a TS in which the para–para CC bond is still too long to have any significant strength. The low activation energy of the reaction is apparently due to a reduction in the repulsion between the two positive charges, which are initially largely localized on the nitrogen atoms. At the time of writing, this work had not yet been published.

Conclusions. The work summarized here has led to a complete revision of previously accepted ideas concerning the mechanisms of pericyclic reactions.

1. Whether a pericyclic reaction is "allowed" or "forbidden" depends, as Evans pointed out many years ago, on whether the cyclic TS for the corresponding synchronous reaction is aromatic or antiaromatic.

2. An antiaromatic pericyclic reaction involves a HOMO–LUMO crossing, regardless of the route followed. The property determining whether orbitals cross is not yet

clearly defined. It is not in any way connected with molecular symmetry.

3. Antiaromaticity can be outweighed by other factors. In such cases, "forbidden" reactions may be synchronous.

4. Previous claims that pericyclic reactions involve "a new principle of bonding" are therefore incorrect. Aromaticity is one of the oldest concepts in organic chemistry.

5. However, a new principle is involved in these reactions, namely the tendency of multibond reactions not to take place synchronously (the multibond rule). As a result, supposedly "allowed" reactions may not take place synchronously.

6. The new picture that emerges from all this is both simpler and more precise than the one in terms of "orbital symmetry" still given in current textbooks. I hope for the sake of students that the textbooks will soon change.

Nucleophilic Substitution. Another significant contribution has come from an extensive study of nucleophilic aliphatic substitution, which has had an amusing history.[125] Many must have been puzzled, as I was, by an anomaly in the simple MO description of the S_N2 reaction. The TS (62) for such a process contains a three-center four-electron bond. The corresponding MOs are derived by interaction of a p AO of the carbon atom with AOs of the entering and leaving groups. This situation is exactly analogous to the π bonds in the allyl anion (63). In the reactants 64, two of the four electrons occupy an AO of the attacking anion, while two form the C—Y bond. This situation parallels that in an allyl anion in which one methylene group has been rotated through 90° (65). Now 63 is lower in energy than 65. Why then is 62 not more stable than 64?

Dougherty had, in fact, suggested[126] a possible solution, on the basis of his observation that halide ions combine very exothermically with alkyl halides in the gas phase. He suggested that the products might be the trigonal bipyramidal species (62) commonly believed to be the TSs in S_N2 reactions, and that these species might be stable intermediates rather than TSs. His suggestion seemed, moreover, to be supported by mass

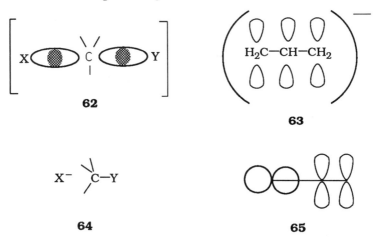

spectrometric evidence that S_N2 reactions commonly take place with little or no activation in the gas phase. If so, as Dougherty pointed out, the activation barriers to such reactions in solution would have to be due entirely to the energy needed to desolvate the halide anion so that the other reactant could approach. This conclusion would have been revolutionary.

According to current theories of chemical reactivity, the difference in energy between the reactants and the TS in a reaction represents the energy needed to bring about the corresponding changes in bonding. The solvent exerts only a secondary role, via differences in solvation energy between the reactants and the corresponding TS. According to Dougherty's suggestion, the solvent is responsible for the *whole* of the activation barriers of S_N2 reactions in solution.

Subsequent experimental work and theoretical calculations showed, however, that Dougherty's adducts were, in fact, electrostatic charge–dipole (CD) complexes with the trigonal bipyramidal species as the TS for interconversion of the CD complexes formed by the reactants and products. A very extensive MNDO study[125] of S_N2 reactions not only supported this interpretation but also accounted for the activation parameters observed in solution. So we were left once again with the paradox.

One could, of course, dismiss this qualitative argument as an indication of the inadequacy of simple-minded MO theory.

However, as time goes on, I become more and more convinced that such "anomalies" simply reflect one's own lack of insight. Simple MO theory always works! A good example is provided by another long-standing "anomaly", namely the apparently unreasonable stability of cyclopropane (66). The strain energy of 66 is no greater than that of cyclobutane (67), even though the angle strain at each vertex in 66 (49.5°) is much greater than that (19.5°) in 67. Indeed, if CCC bending follows a parabolic potential, the strain energy of 66 should be 5 *times* that of 67. This result was recently explained[123] in terms of a fact long known but ignored; that is, that the resonance integral between two hybrid AOs of a given atom does not vanish. A similar argument also accounts for the hitherto unexplained success of our early treatment of bond localization.[30]

When the interactions between the CC bonds in 66 are taken into account, the bonds are seen to form a cyclic σ-conjugated system analogous to the π system in benzene. Thus 66 is aromatic, a surprising conclusion, but one that also accounts for its other anomalous properties.[127] A similar analysis[127] also explained a number of other observations that had been difficult to explain in terms of simple MO theory, in particular the pyramidal structure of the *t*-butyl radical, the bent structure of triplet carbene, and various conformational effects, in particular the anomeric effect.

Returning to the S_N2 reaction, the clue was provided by an MNDO study[125] of the S_N2' reaction of chloride ion with allyl chloride. This reaction was predicted to lead exothermically and without activation to a stable "nonclassical" adduct (68) with a five-center six-electron bond analogous to the π bonds in a benzene Meisenheimer adduct (69). So the S_N2' reaction is predicted to behave in the way expected from simple MO theory! The only reasonable explanation is that carbon is too small to form five unstrained covalent bonds. This problem arises in an S_N2 TS where five groups are attached to the central

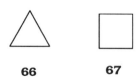

66   **67**

carbon atom, but not in an S_N2' TS (**68**) where each carbon has no more than four neighbors.

If this explanation is correct, the S_N2 reaction might also behave "normally" in the case of a larger central atom. Indeed, calculations[128] for a number of reactions of anions with halosilanes predicted them to lead very exothermically and without activation to trigonal bipyramidal adducts. The fact that "saturated" silicon compounds are far more reactive than carbon analogues has therefore nothing to do with d AOs, a conclusion that now seems to be generally accepted by silicon chemists. The differences between the behavior of heavier elements and that of second-period elements have thus been viewed upside-down. The second-period elements are the ones that behave anomalously because they are too small to accommodate five adjacent groups or pairs of unshared electrons. This, for example, is the reason that nitrogen does not form a pentafluoride.

Nucleophilic substitution at unsaturated carbon does not encounter this difficulty (*cf.* **68**). As expected, MNDO and AM1 agreed[129] in predicting the addition of anions (e.g., HO$^-$) to carboxylic acid derivatives (e.g., esters or amides) to take place very exothermically (ΔH, 20–30 kcal/mol) and without activation. Thus the barriers to such reactions in solution are indeed due entirely to desolvation. This idea so shocked referees of our manuscript, when we sent it to the *Journal of the American Chemical Society*, that we eventually had to publish it elsewhere. However, given that our suggestion has been confirmed by two subsequent high-level ab initio calculations,[130,131] there can now be little doubt that it is correct. Thus the situation envisaged by Dougherty does exist, even if not in the case he originally proposed.

Reactions in solution can thus be divided into two classes,[129] normal *intrinsic barrier* (IB) reactions where the activa-

$$(Cl\cdots CH_2 \doteq CH \doteq CH_2 \cdots Cl)^-$$

68

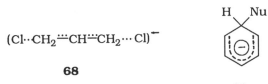

69

tion barriers are due to changes in bonding, and *solvent barrier* (SB) reactions where the barriers are due to desolvation. As indicated, this conclusion has major implications for solution chemistry. It accounts immediately for the well-known differences between nucleophilic substitutions at carbonyl carbon and at saturated carbon. These differences were previously explained in terms of the hard—soft acid—base (HSAB) theory. Our work indicated that they are caused by a difference in mechanism, not by a mystical distinction between different kinds of acids and bases. It seems likely that the same situation may hold in other cases where HSAB theory has been invoked.

Enzyme Reactions. The most significant deduction relates, however, to enzyme reactions.[132,133] No satisfactory explanation had previously been given for their very high reaction rates or for the fact that the high rates apply only to very specific types of compounds, the proper substrates of the enzymes in question. Previous explanations had been based on ad hoc assumptions or a postulated cooperation of different factors in a manner for which there was no other analogy. Our work led to a simple solution of this problem.

The active site of an enzyme consists of a cleft or hollow in which the substrate is adsorbed. It is generally agreed that the proper substrates of enzymes fit their active sites closely, this being a major factor in the specificity of enzyme reactions. Adsorption of such a species must therefore lead to expulsion of all irrelevant water between the enzyme and substrate. Any subsequent reaction between the enzyme and substrate consequently takes place in the absence of water, and follows the same pattern as an analogous reaction in the gas phase. If the reaction in question is an SB one, it should then take place with correspondingly less activation in the enzyme. Alkaline hydrolysis of an amide should, for example, take place without any activation whatsoever.

Removal of the activation barrier would lead to an increase in rate far greater than that brought about by typical proteases (e.g., chymotrypsin). Furthermore, this huge acceleration will apply only to substrates that fit the active site closely and are also adsorbed strongly enough in it to exclude all water.

Species that fail to get rid of the water completely will react much more slowly because the water left behind in the active site acts as an effective inhibitor. In this way, the activity and specificity of enzymes can be explained immediately, without any need for special pleading or special assumptions.

There is, however, a further point to be considered. If an ion were present initially in the substrate or active site, it would be solvated, and its desolvation would require as much energy as desolvation of an analogous ion in solution. If an enzyme reaction involves an ionic reagent, this ionic reagent must then be generated *after* the substrate is adsorbed by some reaction taking place inside the enzyme. Such a process is indeed known to be involved in the hydrolysis of peptides by chymotrypsin.

Chymotrypsin cleaves amides by nucleophilic attack by an alkoxide ion, generated by deprotonation of the hydroxyl group of a serine moiety in the active site. The deprotonation is brought about by the imidazole nitrogen of an adjacent histidine unit and is assisted by the carboxylate group of an aspartate ion buried inside the enzyme. Our model calculations[129] showed the proton transfer to be endothermic by about 30 kcal/mol. The active site is thus initially neutral, allowing the peptide to be adsorbed and water excluded. The exothermic (about 20 kcal/mol) addition of serinoxide ion to the amide carbonyl then acts as a driving force for the proton transfer. The activation energy for the overall reaction should then be about 10 kcal/mol plus any additional activation involved in the proton transfers. The proton transfers, however, probably involve little additional energy, given that they can take place by tunneling. The observed rate of peptide cleavage by chymotrypsin corresponds, in fact, to an effective barrier of 12–13 kcal/mol.

This interpretation[132,133] of enzyme reactions has aroused much interest and rather strong reactions. Five hundred reprints of the original paper vanished in a few weeks! Because of their implications, the conclusions could hardly be accepted by enzyme chemists without a fight. Enzyme reactions had hitherto been regarded as analogues of reactions in solution and had been interpreted accordingly. If we are right, they should be regarded as analogues of reactions in the gas phase. Gas-phase chemistry often differs greatly from solution chemistry. In solution, for example, toluene is a weaker acid than water by about 25 pK units. In the gas phase toluene is a much stronger acid. Acceptance of our suggestion would mean that the whole area of enzyme mechanisms would have to be reevaluated.

Because it is difficult to see any flaw in our arguments, enzyme chemists will presumably have to accept them in the end. It will be interesting to see how this capitulation takes place. I suspect that our calculations will be repeated by some respectable enzyme theorist, without adequate acknowledgment, whereupon he or she will get credit for the idea. There are now a number of cases where ab initioists have in this way gained credit for our contributions.

Refutation of Frontier Orbital Theory

Another contribution that will undoubtedly also arouse strong reactions was a refutation of the frontier orbital (FO) theory. The growing acceptance of FO theory in recent years is a surprising phenomenon because FO theory has no real theoretical basis, because it commonly fails, because it needs a knowledge of orbital coefficients that can be obtained only by using a computer, and because there are no problems that cannot be treated more simply and more effectively by other simpler procedures. FO theory, like PMO theory, is based on the perturbational approach introduced by Coulson and Longuet-Higgins.[25] However, while PMO theory involves only mathematically justifiable approximations, FO theory rests on one which is quite unjustifiable, namely, the neglect of all terms in the expression for a second-order perturbation other than the first. This approximation would be justifiable only if the first term were always much the largest, but this is not the case.

It is therefore not surprising that FO theory frequently fails. Although failures of this kind have been recognized, they have been treated as special cases and have been explained away by making what were, in essence, ad hoc assumptions. The justification for doing this has been that FO theory provided the only satisfactory explanation of a number of phenomena, in particular the course of cycloaddition reactions. The difficulties encountered were, however, due simply to the assumption that "allowed" pericyclic reactions are necessarily synchronous, as predicted by the Woodward–Hoffmann rules. Our work has shown that this is not necessarily the case, and that cycloadditions are among the exceptions. Once it is recognized that the transition states of such reactions are very unsymmetrical, differing little from the biradicals derived by breaking the weaker of the two forming bonds, the rates and regioselectivities of such reactions can be explained simply and reliably by using PMO theory, or even resonance theory. Furthermore, PMO theory and resonance theory still succeed in the not infrequent cases where FO theory fails.

Failures of FO theory are not confined to special cases such as the Diels–Alder reaction or dipolar additions. FO theory also fails to explain quite basic features of such standard reactions as aromatic and aliphatic substitution. One striking example was mentioned earlier, namely electrophilic substitution in 10,9-borazarophenanthrene (**39**). However, the failure of FO theory is not confined to a few exotic cases. It also fails to reproduce the relative reactivities of quite ordinary aromatic compounds, even those of hydrocarbons, and similar disastrous failures appear in applications to S_N2 reactions. In all these cases, PMO theory gives correct predictions. There are, on the other hand, no cases where FO theory succeeds and PMO theory fails. This is not surprising, given that PMO theory has a valid basis in quantum mechanics, whereas FO theory does not. Remember too, that FO theory, unlike PMO theory, requires a knowledge of AO coefficients that can be obtained only by using a computer.

Why then has FO theory gained such wide acceptance when PMO theory is still little-known? The answer is simple: good publicity! I have never been good at publicizing what I have done, a failing I probably inherited from Robinson. Anyway, a general account of the failings of FO theory has just appeared.[134]

Elimination Reactions

Another project involved a very detailed study of bimolecular nucleophilic (E2) elimination reactions.[135] Our calculations reproduced the familiar distinction between eliminations involving neutral molecules (e.g., alkyl halides), which follow Saytzeff's rule, and eliminations involving onium salts, which follow Hoffmann's rule, and they also accounted for the effects of substituents on the rates of elimination. The predicted mechanisms differed, however, from those currently accepted. Thus whereas onium eliminations are usually assumed to take place via TSs similar to those in E1cb eliminations, the main change corresponding to deprotonation of the onium ion by base, AM1 predicted such reactions to be close to the E1 extreme, the main change being weakening of the bond to the leaving (onium) group. Conversely, AM1 predicted the Saytzeff-type eliminations to be E1cb-like processes whereas it has been commonly assumed that they correspond to synchronous E2 reactions.

The discrepancy may well reflect genuine changes in mechanism due to the large solvation energies of ions. Solvation can lead to major differences between reactions in the gas phase and in solution, as our work[125] on nucleophilic substitution shows. However, if this is the case here, the current qualitative theories of chemical reactivity given in textbooks will require revision because they are based on the assumption that solvation plays a relatively minor role. Thus the conventional explanations of E2 reactions imply that Hoffmann-type elimination is to be expected in reactions of E2/E1cb type, even in the absence of solvent. However, our calculations predict Saytzeff-type elimination in such reactions. Further experimental studies of E2 reactions, both in the gas phase and in solution, would clearly be of interest.

Singlet Oxygen

Singlet oxygen, the excited state of O_2 that lies ca. 1 eV above the normal triplet ground state, has been the subject of much interest in recent years because of its involvement in photobiology and because of the theoretical problems it presents.

At first sight, singlet oxygen seems a very "normal" molecule. It can be represented by an apparently normal valence structure (O=O), and its reactions with olefins are apparently consistent with this formulation, the products

corresponding to ene reactions or cycloadditions. It was naturally assumed in the past that the products must be formed in this
way. The analogy is, however, misleading because singlet oxygen does not contain a normal O=O double bond. The bonding involves not two but *six* electrons, four of which fill the
bonding π_x and π_y MOs, while the other two occupy the
corresponding antibonding MOs. The bond between the oxygen atoms is thus not a double bond but a 3–1 bond, consisting
of six bonding electrons and two antibonding electrons, the
former giving rise to a triple bond and the latter to an antibond
that in effect cancels one-third of the triple bond (*see* **70**). Furthermore, because the two antibonding electrons have two MOs
to choose from, oxygen has a triplet ground state in which the
two antibonding electrons occupy different MOs with parallel
spins.

Triplet O_2, with two singly occupied MOs, behaves as a
biradical. Singlet oxygen should, however, act as an electro-
phile, using the empty antibonding MO as an acceptor. Furthermore, because this MO has a node bisecting the OO bond, orbitals of other molecules can overlap with only one or the other
end of the LUMO, not with both. If the other reactant is an
olefin, the adduct should then be a π complex (**71**), the filled π
MO of the olefin acting as the donor. According to π complex
theory, species of this type can be stabilized by back-
coordination if the apical atom or group (here O_2) has a filled
orbital able to interact with the empty antibonding π MO of the
olefin. The filled antibonding π MO of O_2 in **71** meets this condition; *see* **72**. Furthermore, as we saw earlier, π complexes with
back-coordination can be represented equivalently in terms of
three-membered rings, and the same is true here; *see* **73**. There
should be a continuous transition between species of this type

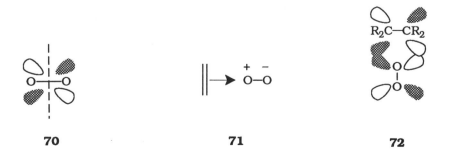

70   **71**   **72**

that are best represented as π complexes with back-coordination (74) and ones best represented as perepoxides (73). If these arguments are correct, the observed products from such reactions (ene products or cycloadducts) must be formed by secondary rearrangements of intermediate perepoxides or π complexes.

Similar comments should also apply to the formation of apparent Diels–Alder products from 1,3-dienes and singlet oxygen; see 75. Here again the analogy is illusory. Because the relevant oxygen π MO has a node bissecting the OO bond, the delocalized system in the transition state for synchronous addition would be of anti-Hückel type and consequently antiaromatic. A simple Diels–Alder addition of singlet oxygen to a 1,3-diene would therefore be "forbidden". Indeed, the same argument suggests that direct addition of singlet oxygen to an olefin to form a dioxetane should be an "allowed" process and might be synchronous. Adducts of this type are, however, rarely formed.

Ab initio studies of the reactions of singlet oxygen have not yet been carried out at a high enough level, and problems arise in our semiempirical procedures because neither MNDO nor AM1 gives a satisfactory estimate of the heat of formation of singlet oxygen. This, however, is one of the cases where MINDO/3 outperforms its successors, and some years ago we used it to study[136] the reactions of singlet oxygen with olefins. Our calculations fully supported the conclusions just given, the first step in each reaction being the formation of a perepoxide or π complex. This result was found even in the case of 1,3-butadiene; see 76. We were also able to locate the transition states for the subsequent rearrangements of the intermediates to the observed products, and our calculations also correctly predicted rearrangement to ene products to be easier than rearrangement to dioxetanes. Our conclusions have moreover been fully confirmed by subsequent experimental studies.

73

74

75   **76**

Other Reactions

The reactions just discussed represent only a small part of our work in this area, being ones where our contributions have led to important and unexpected revisions of existing chemical theory. Organic chemists have spent a vast amount of time and effort in studies of the mechanisms of these reactions, using a wide variety of sophisticated experimental procedures developed for the purpose, and their conclusions had been generally accepted as incontrovertible. The fact that we have been able to make so many major revisions is therefore both surprising and striking.

Ab initio methods could in principle have been used in this connection, and high-level ab initio methods are clearly superior to our current semiempirical ones. However, calculations of this type are currently restricted to reactions of very small molecules. The ab initio studies that have been reported for reactions of more complex systems have necessarily used simpler procedures that are no better than ours and yet cost far more. Studies of this type have moreover usually been restricted to the simplest example of the reaction in question, and the calculations have also often been skimped on to save computing time.

The cost differential between our procedures and equivalent ab initio ones is in fact enormous. My group alone has studied more reactions than all the ab initioists put together, with far more chemical effect and at a tiny fraction of the cost. A full account of our work would be out of place here. However, I will compromise by giving a list of our more significant recent studies. For summaries of our earlier work, *see* refs. 92 and 137:

1. rearrangement of benzyl cations to tropylium ions[138]
2. conversion of toluene radical cation to tropylium[139]
3. 1,2-elimination of H_2 from simple organic cations[140]
4. rearrangement of phenylcarbene to cycloheptatetraene[141]
5. thermal decarboxylation of $H_2C=CH-CH_2COOH$[142]
6. conversion of aza-Dewar benzenes to pyridine[143]
7. hydroboration of alkenes and alkynes[144]
8. boron hydrides and borohydride anions[145]
9. Norrish-type-II reaction of butanal[146]
10. radical addition reactions[147]
11. kinetic isotope effects[148]
12. rearrangements of C_4H_7 radicals[149]
13. deoxygenation of carbonyl compounds by atomic carbon[150]
14. a detailed survey of C_4H_4 species and their interconversions[151]
15. cheletropic reactions[152]
16. interconversion of cyclopropyl and cyclobutyl cations[153]
17. reactions of organotin compounds[154]
18. the double proton shift in azomethine[155]
19. rearrangement of azulene to naphthalene[156]
20. thermal rearrangements of $C_{10}H_8$ species[157]
21. tunneling in Jahn–Teller active molecules[158]
22. ring opening in phenyl radical[159]
23. the Reformatsky reaction[160]
24. 1,5-sigmatropic hydrogen shift in 1,3-pentadiene[161]
25. mechanism of the chain extension step in the biosynthesis of fatty acids[162]
26. mode of action of carbonic anhydrase[163]
27. ozonolysis of ethylene and of *cis*- and *trans*-2-butene[164]

I should point out that this list refers only to the relatively small part of our work that dealt with theoretical studies of reactions. The rest was concerned with experimental studies, with the development of new theoretical procedures, and with calculations of molecular properties.

Most of the examples listed refer to systems that could not have been studied in comparable detail by any adequate ab initio procedure. Furthermore, most of them involved calculations for a number of different individual examples of the reaction in question, whereas analogous ab initio studies are usually confined to one, namely the simplest. The most striking case is that described in ref. 157, a study of two rather remarkable reactions, involving the thermal rearrangement of azulene to naphthalene and the thermal scrambling of carbon atoms in naphthalene. This investigation involved calculations for more than 30 different $C_{10}H_8$ isomers and location of the transition states for their interconversions. Because the reactions involved biradical-like intermediates, an ab initio study would have had to be carried out using a large basis set (at least 6-31G*), together with allowance for electron correlation. Such a calculation for *one* of these reactions would be a major undertaking. Our results, in conjunction with experiment, seem incidentally to have established the mechanisms of both processes.

High-Temperature Superconductors

Some years ago, when I was in my 60s, I was described in print as "the *enfant terrible* of theoretical chemistry". The last contribution I will mention, of still uncertain value, shows that I am still not afraid to stick my neck out. This contribution concerns a possible new mechanism for superconductivity that may operate in the new high-temperature superconductors that were discovered a few years ago. In spite of heroic efforts by all the experts in the field, nobody has yet been able to account for their behavior. So unless this is an area where the "experts" are singularly inexpert, superconductivity in these new materials must involve some entirely new mechanism.

Three years ago, I suggested a possible explanation[165] based on the idea that electrical conduction in these materials takes place by an electrochemical mechanism involving electron hopping along a chain of metal atoms that can exist in different valence states rather than by the usual band mechanism that operates in metals. The electron hopping is mediated by lattice vibrations that cooperate to bring about a unidirectional flow of charge. My original communication gave only the germ of the idea, which I have since developed in more detail. However, I have not yet been able to publish it because of negative opinions by the experts on superconductivity to whom it has been sent to referee. The trouble here is that the experts in the area are physicists who expect theories to be expressed in mathematical terms, using the terminology of physics. Chemical theories, on the other hand, are stated in qualitative terms using chemical terminology, and my theory is no exception. The referees have therefore not only failed to understand it but have also assumed that it cannot possibly contain any new ideas of real value. Because my suggested mechanism bears a superficial resemblance to one that the physicists have refuted, referees naturally assume that the two are the same, which they are not. One referee was indeed kind enough to express surprise that a mere chemist should have come up with the idea independently! I am still trying because I want to know whether I am right and because I lack the theoretical skills needed to check its validity. So far, none of the chemists who have read my manuscript have found any obvious holes in it. However, regardless of whether my suggestions are a lot of nonsense, this episode provides rather striking support for the comments I made earlier about the effects of the compartmentalization of science. All the experts who have been consulted are sure that my theory must be by definition wrong, even though they have not understood it and even though they have not been able themselves to provide a satisfactory alternative.

End of an Era—The Move to Florida

As my list of publications and the examples described show, my group did a lot of good work in Austin, and while Norman Hackerman was president, Mary and I really enjoyed every minute there. Things unfortunately changed for the worse after he left, as I have already mentioned, and under Peter Flawn's successor the situation finally began to wear me down. I could have kept my chair at UT indefinitely because the Texas legislature had abolished retirement by age at universities in Texas. However, I had had enough. We decided to move to Gainesville, Florida. We knew Gainesville well because I had been consulting for a pharmaceutical company, Pharmatec, which was set up by a leading entrepreneur with Nicholas Bodor in charge to exploit Nicholas's discoveries, which are rather striking. Nicholas, an ex-member of my group, is now graduate research professor of medicinal chemistry at the University of Florida. Alan Katritzky, the Kenan Professor of organic chemistry at the University of Florida (UF), is also an old friend of ours, dating back to our days in England when he was the professor of chemistry at the University of East Anglia. We also knew a lot of the other chemists at UF, in particular the large and active group of theoreticians in the quantum theory project. However, to tell the truth, I was really intending to retire and had been looking forward to it. I have been wanting to do many things for years but never had the time. We found a house that was just about perfect for us and had in fact already moved into it before we left Austin. However, when UF heard I was coming, they offered me a research professorship, so I have had to put off retirement for a while longer.

Our only real regret in leaving Austin was losing Bonnie George. Bonnie was not only by a very wide margin the best secretary I have ever had or come across but also a very good friend to both Mary and me. She was almost incredibly competent. When I left, I had succeeded in promoting her to the highest administrative level I could, but this was really quite inadequate. She not only ran my group, including myself, with incredible efficiency but also was an expert draftsman—or I suppose I should say, draftsperson. She drew all the formulae and

The mainstay of my group in Austin, Texas, in the 1980s, Bonnie George.

figures for our papers. And on top of that she typed one of Mary's books, a critical edition of a classic 16th-century English text, *De Republica Anglorum*. I doubt if anyone else in the world could both type 16th-century English and also draw diagrams for *JACS*! And certainly nobody else could have kept me organized in the way she did. Here in Gainesville, I have already lapsed into chaos.

I should also add that I never had any problems with the chemistry department at UT, which was always wholly supportive and where we had, and have, some very close friends. Indeed, to celebrate my 70th birthday, the department organized an international symposium on physical organic chemistry, which proved to be a really major chemical event. More than 200 chemists from all over the world came to it, in addition to 98 ex-members of my research group, some of whom we had not seen for 30 years, and the papers presented were published in a special issue of *Tetrahedron*.

The Dewar family at the Dewar symposium in Austin, Texas, 1988: Michael, Mary, Steuart, and Robert. Courtesy Tom Holt, Third Eye Photography, Austin, Texas.

I do not intend having a large research group in Gainesville, and I do not know how long I will want to go on working. However, the chance to continue research, for a while at
least, is welcome because it should allow us to develop an effective theoretical treatment of transition metals, which has been
my major goal for the past 15 years. This will not be achieved
by any ab initio procedure in the foreseeable future because the
molecules in question are far too large. Equally, after 10 years of
effort, we have had to admit defeat in attempts to extend
MNDO or AM1 to transition metals. The formalism for the electron repulsion integrals becomes much too cumbersome. We are
currently working on a fourth-generation successor to AM1 in
which this difficulty will not arise and which we hope will also
avoid some of the remaining weaknesses of AM1. Work on this
new treatment began 2 years ago in Austin, and at the time of
writing (spring 1991) we have already parameterized it for the
organic elements (C, H, O, and N) with satisfactory results.

*With Hans Bock and Edel Wasserman at the Dewar symposium. Hans, profes-
sor of chemistry at the University of Frankfurt, is one of the leading chemists
in Germany and a recognized authority on wine. Ed was indirectly responsible
for introducing me to computing, when he was at Bell Telephone Laboratories
in Murray Hill, New Jersey. He is now associate director of chemical sciences
at E. I. du Pont de Nemours and Company in Wilmington, Delaware. Both
have been close friends for many years.*

Unless unexpected problems arise, we should have results for at least one transition metal within a year. What will this new treatment be called? Clearly not AM2!

So the move to Gainesville has been an unqualified success. Mary and I have not been so happy for a long time as we have been since we came here, and this in spite of an unusual spate of sickness and injuries. I am sure that next year will be even better.

Debts Owed, Lessons Learned

I owe a very great debt to the many very able people, 175 in all, who have been members of my group over the years. I have deliberately refrained from mentioning any of them by name in this account because it would have been invidious, given that I have been able to review only a small part of my work. The only exception, Roly Pettit, was a colleague, one way or another, throughout almost the whole of my research career and the first ex-member of my group to die, another victim of smoking. The rest of us have kept in touch for many years through a theoretically annual but in practice somewhat sporadic newsletter. Mary was responsible for the idea and has also done all the writing. However, number 14, which appeared 2 years ago, will be the last. If I had not had such a wonderful secretary for the past 10 years in Austin, the newsletter would probably have expired earlier. It was hard work keeping track of ex-members of my group and extracting news from them. At the end, I think we had lost just two.

I also owe a great debt to the Air Force Office of Scientific Research (AFOSR), in particular to Amos Horney and his successors, Tony Matuszko and Fred Hedberg, for their imaginative support of my research. Without it, I could not have achieved more than a fraction of what I have done, and I certainly could not have developed any of the semiempirical procedures that are now being so widely used by organic chemists.

This support began many years ago in Austin. At that time, I had seven grants from various agencies for different projects, and my research was hampered by the need to fit new work into one or another of the proposed programs and by the time spent writing reports and research proposals. One day I had a phone call from Amos, asking if I was going to be in Washington in the near future. When I said no, he asked if he could come and visit me in Austin. When I said, "Of course," he said, "What about tomorrow?" So he flew down the next morning and explained that instead of using AFOSR's funds to support small research projects, he was planning to use part of them to support completely the research of four individual chemists, the idea being that they would be free to work on anything they liked. He asked if I was prepared to be one of the four. I had already been somewhat surprised that the director of the chemical branch of AFOSR should be willing to fly from Washington to Austin (a long journey in those pre-jet

With Anthony Matuszko at the Dewar symposium in Austin, Texas, in 1988. I owe much to Tony and AFOSR. Most of the work I have done in the past 25 years would have been impossible without their support and encouragement.

days) just to talk to me, and Amos's invitation was even more of a shock. However, I rallied enough to say yes, and for many years, all my research was indeed supported by AFOSR, apart from a small grant from the Robert A. Welch Foundation (which endowed my chair at UT).

I was in fact the only one of the four whom AFOSR continued to support in this way. The effects are clear from my list of publications. My work blossomed forth in all directions because I was free to do anything that interested me, without wondering how to justify its inclusion in one or another of the rather specific programs for which I had previously had grants. Because I have always had very wide interests, this freedom was just wonderful. After some years, with the dollar depreciating, there came a time when AFOSR could no longer support all my research. However, AFOSR continued to be my major source of research funding, especially for my theoretical work, and without it I could not possibly have done the things I have done. I am, and always will be, deeply grateful to Amos Horney, Tony Matuszko, and now Fred Hedberg. I am glad that the work by my group has justified their confidence in me.

I have avoided any mention here of awards, apart from one that was accidentally relevant. While it is very pleasing to get these, as an indication of the regard of one's colleagues for one's work, they are really not important. I do, however, have special feelings about three I have received. I was very happy to get the Kharasch Visiting Professorship at the University of Chicago, and still more to be later chosen as the first recipient of their Wheland Medal. Because our time at Chicago meant a lot to Mary and me, I was happy to be assured that we had left with no hard feelings. The other was the Royal Society's Davy Medal; this was of course a major honor, but the circumstances under which I got it mattered far more to me.

Mary and I have felt for many years that we were really Americans. We have very much enjoyed our time here. We have come to love America and all that it stands for, and America has treated us and our sons well. All our loyalties have lain here for many years. However, we did not, until quite recently, become American citizens, because we had parents in England who lived to be very old and would have been greatly distressed if we had become American citizens while they were alive. When they were all dead, we became citizens, and not long afterward the Royal Society awarded me the Davy Medal. I was happy to know that my colleagues in England did not hold my defection against me.

What lessons can be drawn from my career? Undoubtedly the most important is the need to publicize one's work. In the past, this was not necessary. Indeed, attempts to do so were liable to backfire. With the present prevalence of operators, the situation has changed. Now, unless one blows one's own trumpet *fortissimo*, people tend to assume that one has no reason for blowing it at all.

A second lesson is the disastrous effect of the growing urge for conformity in chemistry. This trend has been made worse by the tendency to subdivide chemistry into smaller and smaller minidisciplines. The easiest way for a young chemist to succeed nowadays is to produce minor advances in one of these areas, with, of course, full acknowledgement of the contributions of the accepted authorities in it. Any attempt at *real* innovation is extremely dangerous. I was lucky enough to have already made my reputation before the situation became intolerable. I would not like to be starting on a career in chemistry today.

Many of these minikingdoms relate to areas of little current chemical significance. Such an area does not attract good chemists or good students. The "experts" in such an area are especially hostile to intrusions from outside. This hostility has a stultifying effect if the area later becomes important. Because people are unwilling to challenge the expertise of their colleagues, or of experts in other areas, the Old Guard can continue for many years to obstruct innovations they dislike. All experts may be equal, but some are more equal than others!

The damage has been greater in America than elsewhere because of a disastrous change about 15 years ago in the organization of the National Science Foundation (NSF). Decisions concerning NSF research proposals were formerly made by a committee of outside experts in each discipline, a system that worked very well and is still used by the other Federal agencies. The members of such a committee had the authority, ability, and standing to form their own opinions of research proposals, and they usually did so in a very responsible manner. The decisions are now made "in-house" by program directors with little freedom to do other than follow the advice of the referees. One strongly negative report is usually enough to prevent funding of a proposal. Proposals opposed by the experts in any area consequently have little chance of being funded.

Because this system has now been in operation for many years, chemists seem to assume that it developed, from experi-

Michael today.

ence, as the best way to allocate research funds. I happen to know the true story behind it. I was asked to serve on the NSF chemistry committee just after the change had been made. Not knowing that it had been made, I accepted, feeling a moral responsibility to play my part in what had previously been a very important function. When I discovered that the committee no longer dealt with research proposals, I naturally withdrew my acceptance. Kent Wilson, who was then the director of NSF, was concerned and asked me to visit him in Washington. When I asked him why the change had been made, he told me that he felt the morale of the NSF staff was low and might be improved if they were given the responsibility of assessing research proposals! There was never any suggestion that the previous system had not been satisfactory. The change was made simply for the supposed benefit of the NSF staff.

The tendency to meddle with things that are running perfectly well is of course a common error. Another example is the editorship of the *Journal of the American Chemical Society*. Previously, the *Journal of the American Chemical Society* had a single editor and editing the journal was his major function. Under Albert Noyes and Marshall Gates, the *Journal of the American Chemical Society* became, by common consent, the world's leading chemical journal. A competent editor in such a position acquires a sufficient overall knowledge of chemistry and chemists to make his own assessment of controversial papers.

The change at the *Journal of the American Chemical Society* from a single permanent editor to a group of temporary part-time editors has had an effect similar to that of the change at NSF. The present editors do not have the time to read papers themselves because their time is taken up with their own research. For the same reason, they have neither the time nor the inclination to try to extend their chemical interests beyond their own limited areas. They are therefore forced to rely on the opinions of the official "experts" whom they naturally choose as referees. Whereas the *Journal of the American Chemical Society* used to provide an effective and valuable forum for new ideas, it is now virtually impossible to publish anything there that runs counter to party lines.

I have run into problems of this kind more often than most other people because of my chemical upbringing. In my time, there was no real distinction between the different areas of organic chemistry. Robinson was not only one of the leading theoretical organic chemists of the time but also a leader in natural product chemistry and synthesis. To me, quantum mechanical calculations represent no more than a tool for studying chemistry by providing, in effect, a new kind of experimental technique to be used in the same way as, and in conjunction with, all the other chemical techniques (e.g., kinetic studies or NMR or mass spectrometry). My own work has covered very wide areas of chemistry, involving both experiment and theory. As I mentioned earlier, much of my work has, as a result, involved experiments designed to solve specific problems in a wide variety of areas. This is exactly the kind of work likely to antagonize local "experts". Because I have also never been a respecter of persons, let alone of "experts", my contributions have also not always been expressed too tactfully.

My theoretical work has been the best (or worst) example, for reasons which by now should be fairly obvious. I could never have done this work if I had had to rely on NSF for support. Even now, when our procedures are being used by chemists all over the world, and when their value is recognized by the leading theoreticians, an occasional ab initio referee still dismisses them as "worthless"!

Another perennial problem is the tendency of chemists to take chemistry too seriously. Chemistry is something one should enjoy. Too many chemists nowadays seem to treat it as a kind of religion, regarding criticism of their chemical beliefs with the same outrage that a Catholic would regard criticism of the doctrine of the virgin birth. People seem to have forgotten that the first motive of any scientist should be curiosity. One should be trying to find out how nature works, trying to find the answers to questions about it, because one wants to know the answers, not for any other reason. Equally, the object in any scientific discussion should be to find out who is right, not blind defense of preconceived ideas.

I was reared in the Oxford tradition of medieval scholasticism, where the purpose of argument was to test a proposition, not to defend one's ego or beliefs. One defends a proposition to the best of one's ability, within the confines of legitimate argument, to find out if it is true, not because one believes it. If one cannot defend a position, one knows one has to abandon it. I have always been ready, as the record shows, to abandon *any* chemical belief in the face of valid arguments or new evidence. As I have already pointed out, I was indeed the only participant in the notorious norbornyl controversy to change sides, in response to our own calculations and new experimental evidence. Most of my ex-allies would probably have burnt me at the stake, had one been handy. I still have a letter from one of them which begins: "Dear Michael: How can you do this to me after all these years?" The amusing thing was that I had a much greater vested interest than anyone else in the interpretation I abandoned because I first introduced the MO description of nonclassical carbonium ions that made their existence reasonable!

Chemistry is, or should be, fun. I have always found it so,[166] in spite of periodic battles with the "True Believers", and I hope that my many students have got the message.

Eastern cooking (mainly Chinese) has been a major interest of mine for many years.

And the future? Who can say? For many chemical purposes, AM1 is currently the best approach available. However, it is by no means the best that can be achieved by our type of semiempirical approach. It was the best that could be achieved at the time, subject to my basic condition that calculations should be feasible for relatively large molecules, using readily available computers. At the time AM1 was developed, the standard chemical computer was the VAX 11/780. Now that workstations are available that are 30–40 times faster than the VAX and cost far less, similar treatments could be developed using better basic approximations. Indeed, we have almost completed a new treatment of this type ourselves. Because organic chemists all over the world are now beginning to use our procedures routinely, the demand for better ones is likely to become vocal soon.

Given that I won't be around much longer, I would like to see other chemists joining in the work. Although the semiempirical approach seems likely to remain for many years the most promising for practical use in chemistry, the ab initio establishment is clearly unlikely to be interested in developing new semiempirical procedures. It would be a pity if an influx of eager semiempiricists led to the kind of chaotic proliferation that has taken place in the ab initio area, but chemists are likely, if only in self-defense, to keep the situation under control. My other hope is that the objective Boys sought, a basically new ab initio treatment capable of giving chemically accurate results a priori, is achieved soon.

Appendix A

Compound X has the formula $C_6H_6O_3$.

X is insoluble in water but soluble in organic solvents. It can be steam distilled. X dissolves slowly in dilute alkali and remains in solution on acidification. However, it is regenerated by steam distillation of the acidified solution.

X seems to contain a carbonyl group. Thus with semicarbazide ($H_2NNHCONH_2$) it forms a derivative $C_7H_9O_3N_3$.

What is X? For the answer, *see* Appendix C.

Appendix B

The activation energy of a reaction is the difference in energy between the reactants and the transition state (TS). The activation energy can be divided into contributions from (a) changes in energies of localized bonds, (b) the change in delocalization energy, and (c) changes in solvation energy. For a given kind of reaction, for example aromatic substitution, a and c will be approximately the same for all examples. The relative rates, therefore, depend only on the difference in delocalization energy between the reactants and the TS.

The delocalized systems in the reactants and TS usually differ only in that one of them includes an extra AO. Thus, in the case of aromatic substitution, assuming the TS to be the Wheland intermediate, the delocalized system in the TS involves one less atom than the reactant (**77** and **78**). In the case of alternant conjugated systems, these differences in energy can be found by using simple first-order perturbation theory.[48] Thus, for aromatic substitution at atom t in an alternant hydrocarbon (AH), the difference (*localization energy*) is given by βN_t, where the *reactivity number* (N_t) is twice the sum of the coefficients (a_i) of AOs in the nonbonding MO (NBMO) of the TS at the atoms (i) adjacent to atom t (**70**).

These coefficients can be found very quickly and easily by a "pencil and paper" procedure due to Longuet-Higgins. A similar treatment holds for reactions where the TS contains one conjugated atom more than the reactant (e.g. solvolysis of benzyl chloride, **79**→**80**), and the effect of heteroatoms can also be predicted by a similar and equally simple first-order treatment. (Details are given in ref. 48).

$$N_t = 2 (a_r + a_s)$$

Appendix C

The compound is dimethylmaleic anhydride! It is unique in the ease with which it is regenerated from the corresponding acid. It also forms imides easily with typical carbonyl reagents. Because formation of such imides involves loss of a molecule of water, it mimics formation of a normal carbonyl derivative.

References

1. Robinson, R. *Outline of an Electrochemical (Electronic) Theory of the Course of Organic Reactions*; Institute of Chemistry: London, 1932.

2. Dewar, M. J. S.; King, F. E. *J. Chem. Soc.* **1945,** 114.

3. Dewar, M. J. S. *J. Chem. Soc.* **1944,** 615, 619.

4. Cornforth, J. S.; Cornforth, R. H.; Dewar, M. J. S. *Nature (London)* **1944,** 153, 317.

5. Dewar, M. J. S. *Nature (London)* **1945,** 155, 50.

6. Dewar, M. J. S. *Nature (London)* **1945,** 155, 141.

7. Dewar, M. J. S. *Nature (London)* **1945,** 156, 784.

8. Dewar, M. J. S. *J. Chem. Soc.* **1946,** 406.

9. Dewar, M. J. S. *Discuss. Faraday Soc.* **1947,** 2, 50.

10. Coulson, C. A.; Dewar, M. J. S. *Discuss. Faraday Soc.* **1947,** 2, 54.

11. Nevell, T. P.; de Salas, E.; Wilson, C. L. *J. Chem. Soc.* **1939,** 1188.

12. Dewar, M. J. S. *The Electronic Theory of Organic Chemistry*; Oxford University Press: London, 1949.

13. Dewar, M. J. S. *Bull. Soc. Chim. Fr.* **1951**, *18*, C71.

14. Simonetta, M.; Winstein, S. *J. Am. Chem. Soc.* **1954**, *76*, 18.

15. Dewar, M. J. S.; Marchand, A. P. *Annu. Rev. Phys. Chem.* **1965**, *16*, 321.

16. Dewar, M. J. S.; Ford, G. P. *J. Am. Chem. Soc.* **1979**, *101*, 783.

17. Dewar, M. J. S. *Bull. Soc. Chim. Fr.* **1951**, *18*, 79.

18. For a historical account and references, *see* ref. 16.

19. Dewar, M. J. S. *J. Chem. Soc.* **1946**, 777.

20. Bamford, C. H.; Dewar, M. J. S. *Proc. Roy. Soc. Ser. A* **1948**, *192*, 309.

21. Bamford, C. H.; Dewar, M. J. S. *Nature (London)* **1949**, *163*, 215.

22. Bamford, C. H.; Dewar, M. J. S. *J. Soc. Dyers Colour.* **1949**, *65*, 674.

23. Bamford, C. H.; Dewar, M. J. S. *J. Chem. Soc.* **1949**, 2877.

24. Dewar, M. J. S.; Longuet-Higgins, H. C. *Proc. Roy. Soc. Ser. A* **1952**, *214*, 482.

25. Coulson, C. A.; Longuet-Higgins, H. C. *Proc. Roy. Soc. Ser. A* **1947**, *191*, 39; *192*, 16; **1948**, *193*, 447, 456; *198*, 188.

26. Coulson, C. A.; Rushbrooke, G. S. *Proc. Cambridge Philos. Soc.* **1940**, *36*, 193.

27. Longuet-Higgins, H. C. *J. Chem. Phys.* **1950**, *74*, 3341, 3345, 3350, 3353, 3357.

28. Dewar, M. J. S. *J. Am. Chem. Soc.* **1952**, *74*, 3341, 3345, 3350, 3353, 3355, 3357.

29. Dewar, M. J. S. *Annu. Rep. Prog. Chem.* **1956**, *53*, 133.

30. Dewar, M. J. S.; Pettit, R. *J. Chem. Soc.* **1954**, 1617.

31. Dewar, M. J. S.; Schmeising, H. N. *Tetrahedron* **1959**, *5*, 166.

32. Dewar, M. J. S.; Schmeising, H. N. *Tetrahedron* **1960**, *11*, 96.

33. Dewar, M. J. S. *Hyperconjugation*; Ronald: New York, 1962.

34. Dewar, M. J. S.; Mole, T.; Warford, E. W. T. *J. Chem. Soc.* **1956,** 3581.

35. Dewar, M. J. S.; Mole, T. *J. Chem. Soc.* **1957,** 342.

36. Dewar, M. J. S.; Maitlis, P. M. *J. Chem. Soc.* **1957,** 2518, 2521.

37. Dewar, M. J. S.; Sampson, R. J. *J. Chem. Soc.* **1957,** 2789, 2946.

38. Dewar, M. J. S.; Pettit, R. *Chem. Ind. (London)* **1955,** 199; *J. Chem. Soc.* **1956,** 2021, 2026.

39. For a partial review, *see* (a) Dewar, M. J. S. *Prog. Boron Chem.* **1964,** *1,* 235; (b) Dewar, M. J. S. *Boron–Nitrogen Chemistry*; Gould, R. F., Ed.; Advances in Chemistry 42; American Chemical Society: Washington, DC, 1964; p 227.

40. Dewar, M. J. S.; Jones, R. *J. Am. Chem. Soc.* **1968,** *90,* 2137.

41. Dewar, M. J. S.; Lucken, E. C.; Whitehead, M. A. *J. Chem. Soc.* **1960,** 2423.

42. Dewar, M. J. S.; Blackman, L. C. F. *J. Chem. Soc.* **1957,** 162, 165, 171.

43. Dewar, M. J. S.; Blackman, L. C. F. *J. Appl. Chem.* **1957,** *7,* 60.

44. Dave, J. S.; Dewar, M. J. S. *J. Chem. Soc.* **1954,** 4616; **1955,** 4707.

45. Dewar, M. J. S.; Lucken, E. A. C. *Chem. Soc. Spec. Pub.* **1958,** *12,* 223.

46. Dewar, M. J. S.; Grisdale, P. J. *J. Am. Chem. Soc.* **1962,** *74,* 3539, 3541, 3546, 3548.

47. Eagles, D. C. Ph.D. Thesis, University of London, 1956.

48. Dewar, M. J. S.; Dougherty, R. C. *The PMO Theory of Organic Chemistry*; Plenum: New York, 1975.

49. Dewar, M. J. S.; Dougherty, R. C.; Fleischer, E. B. *J. Am. Chem. Soc.* **1962,** *84,* 4882.

50. Dewar, M. J. S.; Fahey, R. C. *Angew. Chem., Int. Ed. Engl.* **1964,** *3,* 245.

51. Dewar, M. J. S.; Talati, A. M. *J. Am. Chem. Soc.* **1964**, *86*, 1592.

52. Briegleb, G. (a) *Zwischen-molekule Kräfte und Molekülstructur*; Stuttgart, 1937; (b) *Elektronen-Donator-Acceptor-Komplexe*; Springer-Verlag: Berlin, 1961.

53. (a) Dewar, M. J. S.; Lepley, A. R. *J. Am. Chem. Soc.* **1961**, *83*, 4560; (b) Dewar, M. J. S.; Rogers, H. *J. Am. Chem. Soc.* **1962**, *84*, 395.

54. Dewar, M. J. S.; Sabelli, N. L. *J. Phys. Chem.* **1961**, *34*, 1232; **1962**, *66*, 2310; *Proc. Roy. Soc. Ser. A* **1961**, *264*, 431.

55. Chung, A. L. H.; Dewar, M. J. S. *J. Chem. Phys.* **1965**, *42*, 756.

56. Dewar, M. J. S.; Kubba, V. *Tetrahedron* **1959**, *7*, 213; *J. Org. Chem.* **1960**, *25*, 1222.

57. Dewar, M. J. S.; Jones, R. *J. Am. Chem. Soc.* **1967**, *89*, 2408.

58. Dewar, M. J. S.; Thompson, C. C., Jr. *Tetrahedron Suppl.* **1966**, *7*, 97.

59. Bentley, M. D.; Dewar, M. J. S. *Tetrahedron Lett.* **1967**, 5043.

60. Bentley, M. D.; Dewar, M. J. S. *J. Am. Chem. Soc.* **1968**, *90*, 1075.

61. Dewar, M. J. S.; Schroeder, J. P. *J. Am. Chem. Soc.* **1964**, *86*, 5235; *J. Org. Chem.* **1965**, *30*, 3485.

62. Dewar, M. J. S.; Schroeder, J. P. *J. Org. Chem.* **1965**, *30*, 2296.

63. Dewar, M. J. S.; Patterson, D.; Simpson, W. I. *J. Chem. Soc., Dalton Trans.* **1973**, 2381.

64. Dewar, M. J. S.; Worley, S. D. *J. Chem. Phys.* **1969**, *50*, 654; **1970**, *51*, 263.

65. Bischoff, P. K.; Dewar, M. J. S.; Goodman, D. W.; Jones, T. B. *J. Organomet. Chem.* **1974**, *82*, 89.

66. Dewar, M. J. S.; David, D. E. *J. Am. Chem. Soc.* **1980**, *102*, 7387.

67. Dewar, M. J. S.; Tien, T.-P. *J. Chem. Soc., Chem. Commun.* **1985**, 1243.

68. (a) Dewar, M. J. S.; de Llano, C. *J. Am. Chem. Soc.* **1969**, *91*, 789. (b) Dewar, M. J. S.; Harget, A. J. *Proc. Roy. Soc. Ser. A* **1970**, *315*, 443.

69. Dewar, M. J. S.; Klopman, G. *J. Am. Chem. Soc.* **1967**, *89*, 3089.

70. Dewar, M. J. S.; Baird, N. C. *J. Chem. Phys.* **1969**, *50*, 1262.

71. Bingham, R. C.; Dewar, M. J. S.; Lo, D. H. *J. Am. Chem.Soc.* **1975**, *97*, 1285, 1294, 1302, 1307.

72. Dewar, M. J. S.; Thiel, W. *J. Am. Chem. Soc.* **1977**, *99*, 4899, 4907.

73. Dewar, M. J. S.; Zoebisch, E. G.; Healy, E. F.; Stewart, J. J. P. *J. Am. Chem. Soc.* **1985**, *107*, 3902.

74. Dewar, M. J. S. *J. Mol. Struct.* **1983**, *100*, 41.

75. Roothaan, C. C. J. *Rev. Mod. Phys.* **1951**, *23*, 69.

76. Hall, G. G. *Proc. Roy. Soc. Ser. A* **1951**, *205*, 541.

77. (a) Dewar, M. J. S.; Storch, D. M. *J. Am. Chem. Soc.* **1985**, *107*, 3898. (b) Dewar, M. J. S.; O'Connor, B. M. *Chem. Phys. Lett.* **1986**, *138*, 141.

78. (a) Dewar, M. J. S.; Ford, G. P. *J. Am. Chem. Soc.* **1977**, *99*, 1685. (b) Dewar, M. J. S.; Thiel, W. *J. Am. Chem. Soc.* **1977**, *99*, 4899.

79. Dewar, M. J. S.; Ford, G. P. *J. Am. Chem. Soc.* **1977**, *99*, 7822.

80. Brown, S. B.; Dewar, M. J. S.; Ford, G. P.; Nelson, D. J.; Rzepa, H. S. *J. Am. Chem. Soc.* **1978**, *100*, 7832.

81. Dewar, M. J. S.; Haddon, R. C.; Suck, S. H. *J. Chem. Soc., Chem. Commun.* **1974**, 611.

82. Bergman, J. G.; Dewar, M. J. S.; Suck, S. H.; Weiner, P. K. *Chem. Phys. Lett.* **1976**, *38*, 226, 228.

83. Dewar, M. J. S.; Yamaguchi, Y.; Suck, S. H. *Chem. Phys. Lett.* **1978**, *59*, 541.

84. Dewar, M. J. S.; Lo, D. H. *Chem. Phys. Lett.* **1975**, *33*, 298.

85. Dewar, M. J. S.; Kollmar, H. W.; Suck, S. H. *J. Am. Chem. Soc.* **1975**, *96*, 5569; *97*, 5590.

86. Dewar, M. J. S.; Yamaguchi, Y.; Suck, S. H. *Chem. Phys. Lett.* **1977**, *50*, 175, 259.

87. Dewar, M. J. S.; Rzepa, H. S. *J. Am. Chem. Soc.* **1978**, *100*, 784.

88. Dewar, M. J. S.; Li, W.-K. *J. Am. Chem. Soc.* **1974**, *96*, 5569.

89. Washburn, W. N. *J. Am. Chem. Soc.* **1975**, *97*, 1615.

90. Breslow, R.; Napierski, J.; Clarke, T. C. *J. Am. Chem. Soc.* **1975**, *96*, 6275.

91. Dewar, M. J. S.; Landman, D. *J. Am. Chem. Soc.* **1977**, *99*, 6179.

92. Dewar, M. J. S. *Chem. Br.* **1975**, *11*, 97.

93. Woodward, R. B.; Hoffmann, R. *Angew. Chem., Int. Ed. Engl.* **1969**, *8*, 781.

94. Evans, M. G. *Trans. Faraday Soc.* **1939**, *35*, 824.

95. Hoffmann, R.; Woodward, R. B. *J. Am. Chem. Soc.* **1965**, *87*, 2048.

96. (a) Craig, D. D. *J. Chem. Soc.* **1959**, 997. (b) Heilbronner, E. *Tetrahedron Lett.* **1964**, 1923.

97. (a) Dewar, M. J. S.; Kirschner, S.; Kollmar, H. W. *J. Am. Chem. Soc.* **1974**, *96*, 5240; (b) Dewar, M. J. S.; Kirschner, S.; Kollmar, H. W.; Wade, L. E. *J. Am. Chem. Soc.* **1974**, *96*, 5242; (c) Dewar, M. J. S.; Kirschner, S. *J. Am. Chem. Soc.* **1974**, *96*, 5244.

98. Lischka, H.; Kohler, H.-J. *J. Am. Chem. Soc.* **1978**, *100*, 5297.

99. Dewar, M. J. S.; Reynolds, C. H. *J. Am. Chem. Soc.* **1984**, *106*, 6619.

100. Dewar, M. J. S.; Ruiz, J. M. *Tetrahedron* **1987**, *43*, 2661.

101. Ansell, M. F.; Selleck, M. E. *J. Chem. Soc.* **1936**, 1238.

102. Dewar, M. J. S.; Reynolds, C. H. *J. Am. Chem. Soc.* **1984**, *106*, 1744.

103. Dewar, M. J. S.; Rzepa, H. S. *J. Am. Chem. Soc.* **1977**, *99*, 7432.

104. Cone, C.; Dewar, M. J. S.; Landman, D. *J. Am. Chem. Soc.* **1977,** *99,* 372.

105. Dewar, M. J. S.; Kirschner, S. *J. Chem. Soc., Chem. Commun.* **1975,** 461.

106. Dewar, M. J. S.; Kirschner, S.; Kollmar, H. W. *J. Am. Chem. Soc.* **1974,** *96,* 7579.

107. Dewar, M. J. S. *Angew. Chem., Int. Ed. Engl.* **1971,** *10,* 761.

108. Friedheim, J. E. Ph.D. Dissertation, The University of Texas at Austin, 1986.

109. Dewar, M. J. S. *J. Am. Chem. Soc.* **1984,** *106,* 209.

110. Dewar, M. J. S.; Olivella, S.; Rzepa, H. S. *J. Am. Chem. Soc.* **1978,** *100,* 5650.

111. Firestone, R. A. *J. Org. Chem.* **1972,** *37,* 218.

112. Dewar, M. J. S.; Olivella, S.; Stewart, J. J. P. *J. Am. Chem. Soc.* **1986,** *108,* 5771.

113. Dewar, M. J. S.; Dennington, R. D. to be published.

114. Bernardi, F.; Bottoni, J.; Robb, M. A.; Field, M. J.; Hillier, I. H.; Guest, M. F. *J. Am. Chem. Soc.* **1988,** *110,* 3050.

115. Doering, W. E.; Toscano, V. G.; Beasley, G. H. *Tetrahedron* **1971,** *27,* 299.

116. Dewar, M. J. S.; Wade, L. E. *J. Am. Chem. Soc.* **1973,** *95,* 290; **1977,** *99,* 4417.

117. Dewar, M. J. S.; Ford, G. P.; McKee, M. L.; Rzepa, H. S.; Wade, L. E. *J. Am. Chem. Soc.* **1973,** *95,* 290; **1977,** *99,* 5069.

118. Dewar, M. J. S.; Jie, C. *J. Am. Chem. Soc.* **1987,** *109,* 5893.

119. Doering, W. E.; Roth, W. R. *Tetrahedron* **1962,** *18,* 67.

120. Dewar, M. J. S.; Jie, C. *J. Chem. Soc., Chem. Commun.* **1987,** 1451.

121. Dewar, M. J. S.; Jie, C. *Tetrahedron* **1988,** *44,* 1351; *J. Chem. Soc., Chem. Commun.* **1989,** 98.

122. Doering, W. E.; Troise, C. A. *J. Am. Chem. Soc.* **1985,** *107,* 5739.

123. Dewar, M. J. S.; Healy, E. F. *Chem. Phys. Lett.* **1987,** *141,* 521.

124. Dewar, M. J. S.; Jie, C. *J. Am. Chem. Soc.* **1989,** *111,* 511.

125. Dewar, M. J. S.; Carrion, F. *J. Am. Chem. Soc.* **1984,** *106,* 3531.

126. Dougherty, R. C. *Org. Mass Spectrom.* **1974,** *8,* 85.

127. Dewar, M. J. S. *J. Am. Chem. Soc.* **1984,** *106,* 669.

128. Dewar, M. J. S.; Healy, E. F. *Organometallics* **1982,** *1,* 1705.

129. (a) Dewar, M. J. S.; Storch, D. M. *J. Chem. Soc., Chem. Commun.* **1985,** 94. (b) *J. Chem. Soc., Perkin Trans.* 2 **1989,** 877.

130. Weiner, S. J.; Singh, U. C.; Kollman, P. A. *J. Am. Chem. Soc.* **1985,** *107,* 2219.

131. Ewig, C. S.; Van Wazer, J. R. *J. Am. Chem. Soc.* **1986,** *108,* 4774.

132. Dewar, M. J. S.; Storch, D. M. *Proc. Natl. Acad. Sci. U.S.A.* **1985,** *82,* 2225.

133. Dewar, M. J. S. *Enzyme* **1986,** *36,* 8.

134. Dewar, M. J. S. *THEOCHEM* **1989,** *200,* 301.

135. Dewar, M. J. S.; Yuan, Y.-C. *J. Am. Chem. Soc.* **1990,** *112,* 2088, 2095.

136. Dewar, M. J. S.; Thiel, W. *J. Am. Chem. Soc.* **1975,** *97,* 3978; **1977,** *99,* 2338.

137. (a) Dewar, M. J. S. *J. Chem. Soc., Faraday Discuss.* **1977,** 62. (b) Dewar, M. J. S. In *Further Perspectives in Organic Chemistry*; Ciba Foundation: London, 1978; Symposium 53 (New Series); p 107.

138. (a) Cone, C.; Dewar, M. J. S.; Landman, D. *J. Am. Chem. Soc.* **1977,** *99,* 372. (b) Dewar, M. J. S.; Landman, D. *J. Am. Chem. Soc.* **1977,** *99,* 4633.

139. Dewar, M. J. S.; Landman, D. *J. Am. Chem. Soc.* **1977,** *99,* 2446.

140. Dewar, M. J. S.; Rzepa, H. S. *J. Am. Chem. Soc.* **1977,** *99,* 7432.

141. Dewar, M. J. S.; Landman, D. *J. Am. Chem. Soc.* **1977,** *99,* 6179.

142. Dewar, M. J. S.; Ford, G. P. *J. Am. Chem. Soc.* **1977,** *99,* 8343.

143. Dewar, M. J. S.; Ford, G. P.; Ritchie, J. P.; Rzepa, H. S. *J. Chem. Res.* **1978,** 26.

144. Dewar, M. J. S.; McKee, M. L. (a) *Inorg. Chem.* **1978,** *17,* 1075. (b) *J. Am. Chem. Soc.* **1978,** *100,* 7499.

145. Dewar, M. J. S.; McKee, M. L. *Inorg. Chem.* **1978,** *17,* 1569.

146. Dewar, M. J. S.; Doubleday, C. *J. Am. Chem. Soc.* **1978,** *100,* 4935.

147. Dewar, M. J. S.; Olivella, S.; Rzepa, H. S. *J. Am. Chem. Soc.* **1978,** *100,* 5650.

148. Brown, S. B.; Dewar, M. J. S.; Ford, G. P.; Nelson, D. J.; Rzepa, H. S. *J. Am. Chem. Soc.* **1978,** *100,* 7832.

149. Dewar, M. J. S.; Olivella, S. *J. Am. Chem. Soc.* **1979,** *101,* 4958.

150. Dewar, M. J. S.; Nelson, D. J.; Shevlin, P. B.; Biesiada, K. A. *J. Am. Chem. Soc.* **1981,** *103,* 2802.

151. Dewar, M. J. S.; Carrion, F.; Kollmar, H. W.; Bingham, R. C. *J. Am. Chem. Soc.* **1981,** *103,* 5292.

152. Dewar, M. J. S.; Chantranupong, L. *J. Am. Chem. Soc.* **1983,** *105,* 7152, 7161.

153. Dewar, M. J. S.; Reynolds, C. H. *J. Am. Chem. Soc.* **1984,** *106,* 6388.

154. Dewar, M. J. S.; Grady, G. L.; Kuhn, D. R.; Merz, K. M., Jr. *J. Am. Chem. Soc.* **1984,** *106,* 6773.

155. Dewar, M. J. S.; Merz, K. M., Jr. *THEOCHEM* **1985,** *124,* 183.

156. Dewar, M. J. S.; Merz, K. M., Jr. *J. Am. Chem. Soc.* **1985,** *107,* 6111; **1986,** *108,* 5142.

157. Dewar, M. J. S.; Merz, K. M., Jr. *J. Am. Chem. Soc.* **1986,** *108,* 5146.

158. Dewar, M. J. S.; Merz, K. M., Jr. *J. Phys. Chem.* **1985,** *89,* 4739.

159. Dewar, M. J. S.; Gardiner, W. C., Jr.; Frenklach, M.; Oref, I. *J. Am. Chem. Soc.* **1987,** *109,* 4456.

160. Dewar, M. J. S.; Merz, K. M., Jr. *J. Am. Chem. Soc.* **1987,** *109,* 6553.

161. Dewar, M. J. S.; Healy, E. F.; Ruiz, J. M. *J. Am. Chem. Soc.* **1988,** *110,* 2666.

162. Dewar, M. J. S.; Dieter, K. M. *Biochemistry* **1988,** *22,* 557.

163. Merz, K. M., Jr.; Dewar, M. J. S.; Hoffmann, R. *J. Am. Chem. Soc.* **1989,** *111,* 5636.

164. Dewar, M. J. S.; Hwang, J. *J. Am. Chem. Soc.* **1991,** *113,* 735.

165. Dewar, M. J. S. *Angew. Chem., Int. Ed. Engl.* **1987,** *26,* 1273.

166. *See* the blue pages in the German edition of *Angew. Chem.* for April 1974.

Index

Index

A

A priori procedure
  trouble with referees, 130
  unambiguous predictions,
    129–130
Ab initio and semiempirical
  methods
  cost differential, 167
  reaction complexity, 167, 169
Ab initio calculations
  reliability, 129–130
  very small molecules, 130
Absolute presuppositions, past
  and current ideas, 30–31
Acetic acid, thermal
  decomposition, 53
Activation barrier, enzyme
  reactions, 160–161
Activation energy, relative
  rates, 189
Administrative ability, 67

Air Force Office of Scientific
  Research, research support,
  177–179
All-valence-electron calculations
  computer programs, 126–127
  parameterized versions of ZDO
    approximations, 128
Alternant hydrocarbon
  method, 123
Ansell, Martin, stereochemistry of
  ring formation, 138
Antimalarials, heteroaromatic
  rings, 25–26
Argument
  development of ideas, 29
  Oxford tradition of medieval
    scholasticism, 27–29, 183–184
Arnold, R. T., conference on
  hyperconjugation, 76 (photo)
Aromatic compounds with
  acceptors
  complexes formed, 121–122
  detection procedure, 122

205

Production: Paula M. Befard
Copy Editing and Indexing: Colleen P. Stamm
Acquisition: Robin Giroux

Printed and bound by Maple Press, York, PA

Paper meets minimum requirements of American National Standard
for Information Sciences—Permanence of Paper for Printed Library
Materials, ANSI Z39.48–1984 ∞